Multivariate Analysis of Ecological Data using CANOCO

This book is primarily written for ecologists needing to analyse data resulting from field observations and experiments. It will be particularly useful for students and researchers dealing with complex ecological problems, such as the variation of biotic communities with environmental conditions or the response of biotic communities to experimental manipulation. Following a simple introduction to ordination methods, the text focuses on constrained ordination methods (RDA, CCA) and the use of permutation tests of statistical hypotheses about multivariate data. An overview of classification methods, or modern regression methods (GLM, GAM, loess), is provided and guidance on the correct interpretation of ordination diagrams is given. Seven case studies of varying difficulty help to illustrate the suggested analytical methods, using Canoco for Windows software. The case studies utilize both the descriptive and manipulative approaches, and they are supported by data sets and project files available from the book website.

JAN LEPŠ is Professor of Ecology in the Department of Botany, at the University of South Bohemia, and in the Institute of Entomology at the Czech Academy of Sciences, Czech Republic.

PETR ŠMILAUER is Lecturer in Multivariate Statistics at the University of South Bohemia, Czech Republic.

Multivariate Analysis of Ecological Data using CANOCO

Jan Lepš
University of South Bohemia, and
Czech Academy of Sciences,
Czech Republic

Petr Šmilauer
University of South Bohemia,
Czech Republic

PUBLISHED BY THE PRESS SYNDICATE OF THE UNIVERSITY OF CAMBRIDGE
The Pitt Building, Trumpington Street, Cambridge, United Kingdom

CAMBRIDGE UNIVERSITY PRESS
The Edinburgh Building, Cambridge CB2 2RU, UK
40 West 20th Street, New York, NY 10011-4211, USA
477 Williamstown Road, Port Melbourne, VIC 3207, Australia
Ruiz de Alarcón 13, 28014 Madrid, Spain
Dock House, The Waterfront, Cape Town 8001, South Africa

http://www.cambridge.org

First published 2003
Reprinted 2005

Printed in the United Kingdom at the University Press, Cambridge

Typefaces Lexicon No. 2 10/14 pt and Lexicon No. 1   *System* LaTeX 2_ε [TB]

A catalogue record for this book is available from the British Library

Library of Congress Cataloguing in Publication data

Lepš, Jan 1953–
    Multivariate analysis of ecolological data using CANOCO / Jan Lepš & Petr Šmilauer.
        p.   cm.
    Includes bibliographical references (p.).
    ISBN 0 521 81409 X (hb.)   ISBN 0 521 89108 6 (pb.)
    1. Ecology – Statistical methods.   2. Multivariate analysis.   I. Šmilauer, Petr, 1967-II.
Title.

QH541.15.S72L47 2003
577′.07′27–dc21 2002034801

ISBN 0 521 81409 X hardback
ISBN 0 521 89108 6 paperback

Contents

Preface

The multidimensional data on community composition, properties of individual populations, or properties of environment are the bread and butter of an ecologist's life. They need to be analysed with taking their multidimensionality into account. A reductionist approach of looking at the properties of each variable separately does not work in most cases. The methods for statistical analysis of such data sets fit under the umbrella of 'multivariate statistical methods'.

In this book, we present a hopefully consistent set of approaches to answering many of the questions that an ecologist might have about the studied systems. Nevertheless, we admit that our views are biased to some extent, and we pay limited attention to other less parametric methods, such as the family of non-metric multidimensional scaling (NMDS) algorithms or the group of methods similar to the Mantel test or the ANOSIM method. We do not want to fuel the controversy between proponents of various approaches to analysing multivariate data. We simply claim that the solutions presented are not the only ones possible, but they work for us, as well as many others.

We also give greater emphasis to ordination methods compared to classification approaches, but we do not imply that the classification methods are not useful. Our description of multivariate methods is extended by a short overview of regression analysis, including some of the more recent developments such as generalized additive models.

Our intention is to provide the reader with both the basic understanding of principles of multivariate methods and the skills needed to use those methods in his/her own work. Consequently, all the methods are illustrated by examples. For all of them, we provide the data on our web page (see Appendix A), and for all the analyses carried out by the CANOCO program, we also provide the CANOCO project files containing all the options needed for particular analysis. The seven case studies that conclude the book contain tutorials, where the analysis options are explained and the software use is described. The individual case studies differ intentionally in the depth of explanation of the necessary steps. In the first case study, the tutorial is in a 'cookbook' form, whereas a detailed description of individual steps in the subsequent case studies is only provided for the more complicated and advanced methods that are not described in the preceding tutorial chapters.

The methods discussed in this book are widely used among plant, animal and soil biologists, as well as in the hydrobiology. The slant towards plant community ecology is an inevitable consequence of the research background of both authors.

This handbook provides study materials for the participants of a course regularly taught at our university called Multivariate Analysis of Ecological Data. We hope that this book can also be used for other similar courses, as well as by individual students seeking improvement in their ability to analyse collected data.

We hope that this book provides an easy-to-read supplement to the more exact and detailed publications such as the collection of Cajo Ter Braak's papers and the Canoco for Windows 4.5 manual. In addition to the scope of those publications, this book adds information on classification methods of multivariate data analysis and introduces modern regression methods, which we have found most useful in ecological research.

In some case studies, we needed to compare multivariate methods with their univariate counterparts. The univariate methods are demonstrated using the Statistica for Windows package (version 5.5). We have also used this package to demonstrate multivariate methods not included in the CANOCO program, such as non-metric multidimensional scaling or the methods of cluster analysis. However, all those methods are available in other statistical packages so the readers can hopefully use their favourite statistical package, if different from Statistica. Please note that we have omitted the trademark and registered trademark symbols when referring to commercial software products.

We would like to thank John Birks, Robert Pillsbury and Samara Hamzé for correcting the English used in this textbook. We are grateful to all who read drafts of the manuscript and gave us many useful comments: Cajo Ter Braak, John Birks, Mike Palmer and Marek Rejmánek. Additional useful comments on the text and the language were provided by the students of Oklahoma State University: Jerad Linneman, Jerry Husak, Kris Karsten, Raelene Crandall and Krysten Schuler. Sarah Price did a great job as our copy-editor, and improved the text in countless places.

Camille Flinders, Tomáš Hájek and Milan Štech kindly provided data sets, respectively, for case studies 6, 2 and 7.

P.Š. wants to thank his wife Marie and daughters Marie and Tereza for their continuous support and patience with him.

J.L. insisted on stating that the ordering of authorship is based purely on the alphabetical order of their names. He wants to thank his parents for support and his daughters Anna and Tereza for patience.

Introduction and data manipulation

1.1. Why ordination?

When we investigate variation of plant or animal communities across a range of different environmental conditions, we usually find not only large differences in species composition of the studied communities, but also a certain consistency or predictability of this variation. For example, if we look at the variation of grassland vegetation in a landscape and describe the plant community composition using vegetation samples, then the individual samples can be usually ordered along one, two or three imaginary axes. The change in the vegetation composition is often small as we move our focus from one sample to those nearby on such a hypothetical axis.

This gradual change in the community composition can often be related to differing, but partially overlapping demands of individual species for environmental factors such as the average soil moisture, its fluctuations throughout the season, the ability of species to compete with other ones for the available nutrients and light, etc. If the axes along which we originally ordered the samples can be identified with a particular environmental factor (such as moisture or richness of soil nutrients), we can call them a soil moisture gradient, a nutrient availability gradient, etc. Occasionally, such gradients can be identified in a real landscape, e.g. as a spatial gradient along a slope from a riverbank, with gradually decreasing soil moisture. But more often we can identify such axes along which the plant or animal communities vary in a more or less smooth, predictable way, yet we cannot find them in nature as a visible spatial gradient and neither can we identify them uniquely with a particular measurable environmental factor. In such cases, we speak about **gradients of species composition change**.

The variation in biotic communities can be summarized using one of a wide range of statistical methods, but if we stress the continuity of change

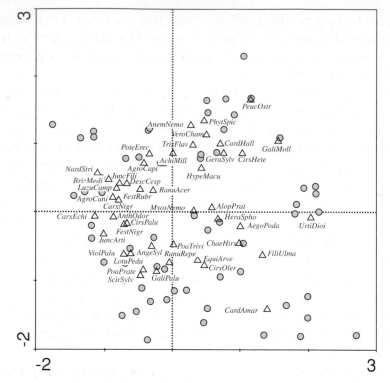

Figure 1-1. Summarizing grassland vegetation composition with ordination: ordination diagram from correspondence analysis.

in community composition, the so-called **ordination methods** are the tools of trade. They have been used by ecologists since the early 1950s, and during their evolution these methods have radiated into a rich and sometimes confusing mixture of various techniques. Their simplest use can be illustrated by the example introduced above. When we collect recordings (samples) representing the species composition of a selected quadrat in a vegetation stand, we can arrange the samples into a table where individual species are represented by columns and individual samples by rows. When we analyse such data with an ordination method (using the approaches described in this book), we can obtain a fairly representative summary of the grassland vegetation using an ordination diagram, such as the one displayed in Figure 1-1.

The rules for reading such ordination diagrams will be discussed thoroughly later on (see Chapter 10), but even without their knowledge we can read much from the diagram, using the idea of continuous change of composition along the gradients (suggested here by the diagram axes) and the idea that **proximity implies similarity**. The individual samples are represented in Figure 1-1 by grey circles. We can expect that two samples that lie near to each other will be much more similar in terms of list of occurring species and even in the relative importance of individual species populations, compared to samples far apart in the diagram.

The triangle symbols represent the individual plant species occurring in the studied type of vegetation (not all species present in the data were included in the diagram). In this example, our knowledge of the ecological properties of the displayed species can aid us in an **ecological interpretation of the gradients** represented by the diagram axes. The species preferring nutrient-rich soils (such as *Urtica dioica*, *Aegopodium podagraria*, or *Filipendula ulmaria*) are located at the right side of the diagram, while the species occurring mostly in soils poor in available nutrients are on the left side (*Viola palustris*, *Carex echinata*, or *Nardus stricta*). The horizontal axis can therefore be informally interpreted as a gradient of nutrient availability, increasing from the left to the right side. Similarly, the species with their points at the bottom of the diagram are from the wetter stands (*Galium palustre*, *Scirpus sylvaticus*, or *Ranunculus repens*) than the species in the upper part of the diagram (such as *Achillea millefolium*, *Trisetum flavescens*, or *Veronica chamaedrys*). The second axis, therefore, represents a gradient of soil moisture.

As you have probably already guessed, the proximity of species symbols (triangles) with respect to a particular sample symbol (a circle) indicates that these species are likely to occur more often and/or with a higher (relative) abundance than the species with symbols more distant from the sample.

Our example study illustrates the most frequent use of ordination methods in community ecology. We can use such an analysis to summarize community patterns and compare the suggested gradients with our independent knowledge of environmental conditions. But we can also test statistically the predictive power of such knowledge; i.e. address the questions such as 'Does the community composition change with the soil moisture or are the identified patterns just a matter of chance?' These analyses can be done with the help of **constrained ordination methods** and their use will be illustrated later in this book.

However, we do not need to stop with such exploratory or simple confirmatory analyses and this is the focus of the rest of the book. The rich toolbox of various types of regression and analysis of variance, including analysis of repeated measurements on permanent sites, analysis of spatially structured data, various types of hierarchical analysis of variance (ANOVA), etc., allows ecologists to address more complex, and often more realistic questions. Given the fact that the populations of different species occupying the same environment often share similar strategies in relation to the environmental factors, it would be very profitable if one could ask similar complex questions for the whole biotic communities. In this book, we demonstrate that this can be done and we show the reader how to do it.

1.2.   Terminology

The terminology for multivariate statistical methods is quite complicated. There are at least two different sets of terminology. One, more general and abstract, contains purely statistical terms applicable across the whole field of science. In this section we give the terms from this set in italics and mostly in parentheses. The other represents a mixture of terms used in ecological statistics with the most typical examples coming from the field of community ecology. This is the set on which we will focus, using the former just to refer to the more general statistical theory. In this way, we use the same terminology as the CANOCO software documentation.

In all cases, we have a data set with the **primary data**. This data set contains records on a collection of observations – **samples** (*sampling units*).* Each sample comprises values for multiple **species** or, less often, the other kinds of descriptors. The primary data can be represented by a rectangular matrix, where the rows typically represent individual samples and the columns represent individual variables (species, chemical or physical properties of the water or soil, etc.).†

Very often our primary data set (containing the *response variables*) is accompanied by another data set containing the *explanatory variables*. If our primary data represent community composition, then the explanatory data set typically contains measurements of the soil or water properties (for the terrestrial or aquatic ecosystems, respectively), a semi-quantitative scoring of human impact, etc. When we use the *explanatory variables* in a model to predict the primary data (like community composition), we might divide them into two different groups. The first group is called, somewhat inappropriately, the **environmental variables** and refers to the variables that are of prime interest (in the role of predictors) in our particular analysis. The other group represents the **covariables** (often referred to as *covariates* in other statistical approaches), which are

---

* There is an inconsistency in the terminology: in classical statistical terminology, **sample** means a collection of sampling units, usually selected at random from the population. In community ecology, sample is usually used for a description of a sampling unit. This usage will be followed in this text. The general statistical packages use the term **case** with the same meaning.

† Note that this arrangement is transposed in comparison with the tables used, for example, in traditional vegetation analyses. The classical vegetation tables have individual taxa represented by rows and the columns represent the individual samples or community types.

also explanatory variables with an acknowledged (or hypothesized) influence on the *response variables*. We want to account for (subtract, partial-out) such an influence **before** focusing on the influence of the variables of prime interest (i.e. the effect of environmental variables).

As an example, let us imagine a situation where we study the effects of soil properties and type of management (hay cutting or pasturing) on the species composition of meadows in a particular area. In one analysis, we might be interested in the effect of soil properties, paying no attention to the management regime. In this analysis, we use the grassland composition as the **species data** (i.e. *primary data set*, with individual plant species as individual *response variables*) and the measured soil properties as the **environmental variables** (*explanatory variables*). Based on the results, we can make conclusions about the preferences of individual plant species' populations for particular environmental gradients, which are described (more or less appropriately) by the measured soil properties. Similarly, we can ask how the management type influences plant composition. In this case, the variables describing the management regime act as environmental variables. Naturally, we might expect that the management also influences the soil properties and this is probably one of the ways in which management acts upon the community composition. Based on such expectation, we may ask about the influence of the management regime **beyond** that mediated through the changes of soil properties. To address such a question, we use the variables describing the management regime as the **environmental variables** and the measured soil properties as the **covariables.**[*]

One of the keys to understanding the terminology used by the CANOCO program is to realize that the data referred to by CANOCO as the **species data** might, in fact, be any kind of data with variables whose values we want to **predict**. For example, if we would like to predict the quantities of various metal ions in river water based on the landscape composition in the catchment area, then the individual ions would represent the individual 'species' in CANOCO terminology. If the **species data** really represent the species composition of a community, we describe the composition using various abundance measures, including counts, frequency estimates, and biomass estimates. Alternatively, we might have information only on the presence or absence of species in individual samples. The quantitative and presence-absence variables may also occur as *explanatory variables*. These various kinds of data values are treated in more detail later in this chapter.

---

[*] This particular example is discussed in the Canoco for Windows manual (Ter Braak & Šmilauer, 2002), section 8.3.1.

Table 1-1. *The types of the statistical models*

Response variable(s)...	Predictor(s)	
	Absent	Present
...**is one**	• distribution summary	• regression models *sensu lato*
...**are many**	• indirect gradient analysis (PCA, DCA, NMDS)	• direct gradient analysis
	• cluster analysis	• discriminant analysis (CVA)

CVA, canonical variate analysis; DCA, detrended correspondence analysis; NMDS, non-metric multidimensional scaling; PCA, principal components analysis.

1.3.   Types of analyses

If we try to describe the behaviour of one or more response variables, the appropriate statistical modelling methodology depends on whether we study each of the response variables separately (or many variables at the same time), and whether we have any explanatory variables (predictors) available when we build the model.

Table 1-1 summarizes the most important statistical methodologies used in these different situations.

If we look at a single response variable and there are no predictors available, then we can only summarize the distributional properties of that variable (e.g. by a histogram, median, standard deviation, inter-quartile range, etc.). In the case of multivariate data, we might use either the ordination approach represented by the methods of **indirect gradient analysis** (most prominent are the principal components analysis – PCA, correspondence analysis – CA, detrended correspondence analysis – DCA, and non-metric multidimensional scaling – NMDS) or we can try to (hierarchically) divide our set of samples into compact distinct groups (methods of cluster analysis, see Chapter 7).

If we have one or more predictors available and we describe values of a single variable, then we use **regression models** in the broad sense, i.e. including both traditional regression methods and methods of analysis of variance (ANOVA) and analysis of covariance (ANOCOV). This group of methods is unified under the so-called **general linear model** and was recently extended and enhanced by the methodology of **generalized linear models (GLM)** and **generalized additive models (GAM)**. Further information on these models is provided in Chapter 8.

If we have predictors for a set of response variables, we can summarize relations between multiple response variables (typically biological species) and one or several predictors using the methods of **direct gradient analysis** (most prominent are redundancy analysis (RDA) and canonical correspondence analysis (CCA), but there are several other methods in this category).

1.4.   Response variables

The data table with response variables* is always part of multivariate analyses. If explanatory variables (see Section 1.5), which may explain the values of the response variables, were not measured, the statistical methods can try to construct hypothetical explanatory variables (groups or gradients).

The response variables (often called species data, based on the typical context of biological community data) can often be measured in a precise (quantitative) way. Examples are the dry weight of the above-ground biomass of plant species, counts of specimens of individual insect species falling into soil traps, or the percentage cover of individual vegetation types in a particular landscape. We can compare different values not only by using the 'greater-than', 'less-than' or 'equal to' expressions, but also using their ratios ('this value is two times higher than the other one').

In other cases, we estimate the values for the primary data on a simple, semi-quantitative scale. Good examples are the various semi-quantitative scales used in recording the composition of plant communities (e.g. original Braun-Blanquet scale or its various modifications). The simplest possible form of data are binary (also called presence-absence or 0/1) data. These data essentially correspond to the list of species present in each of the samples.

If our response variables represent the properties of the chemical or physical environment (e.g. quantified concentrations of ions or more complicated compounds in the water, soil acidity, water temperature, etc.), we usually get quantitative values for them, but with an additional constraint: these characteristics do not share the same units of measurement. This fact precludes the use of some of the ordination methods[†] and dictates the way the variables are standardized if used in the other ordinations (see Section 4.4).

1.5.   Explanatory variables

The explanatory variables (also called predictors or independent variables) represent the knowledge that we have about our samples and that we can use to predict the values of the response variables (e.g. abundance of various

---

* also called dependent variables.
[†] namely correspondence analysis (CA), detrended correspondence analysis (DCA), or canonical correspondence analysis (CCA).

species) in a particular situation. For example, we might try to predict the composition of a plant community based on the soil properties and the type of land management. Note that usually the primary task is not the prediction itself. We try to use 'prediction rules' (derived, most often, from the ordination diagrams) to learn more about the studied organisms or systems.

Predictors can be quantitative variables (concentration of nitrate ions in soil), semi-quantitative estimates (degree of human influence estimated on a 0–3 scale) or factors (nominal or categorical – also categorial – variables). The simplest predictor form is a binary variable, where the presence or absence of a certain feature or event (e.g. vegetation was mown, the sample is located in study area X, etc.) is indicated, respectively, by a 1 or 0 value.

The factors are the natural way of expressing the classification of our samples or subjects: For example, classes of management type for meadows, type of stream for a study of pollution impact on rivers, or an indicator of the presence/absence of a settlement near the sample in question. When using factors in the CANOCO program, we must re-code them into so-called **dummy variables**, sometimes also called **indicator variables** (and, also, binary variables). There is one separate dummy variable for each different value (level) of the factor. If a sample (observation) has a particular value of the factor, then the corresponding dummy variable has the value 1.0 for this sample, and the other dummy variables have a value of 0.0 for the same sample. For example, we might record for each of our samples of grassland vegetation whether it is a pasture, meadow, or abandoned grassland. We need three dummy variables for recording such a factor and their respective values for a meadow are 0.0, 1.0, and 0.0.*

Additionally, this explicit decomposition of factors into dummy variables allows us to create so-called **fuzzy coding**. Using our previous example, we might include in our data set a site that had been used as a hay-cut meadow until the previous year, but was used as pasture in the current year. We can reasonably expect that both types of management influenced the present composition of the plant community. Therefore, we would give values larger than 0.0 and less than 1.0 for both the first and second dummy variables. The important restriction here is that the values must sum to 1.0 (similar to the dummy variables coding normal factors). Unless we can quantify the relative importance of the two management types acting on this site, our best guess is to use values 0.5, 0.5, and 0.0.

---

* In fact, we need only two (generally $K-1$) dummy variables to code uniquely a factor with three (generally K) levels. But the one redundant dummy variable is usually kept in the data, which is advantageous when visualizing the results in ordination diagrams.

If we build a model where we try to predict values of the response variables ('species data') using the explanatory variables ('environmental data'), we often encounter a situation where some of the explanatory variables affect the species data, yet these variables are treated differently: we do not want to interpret their effect, but only want to take this effect into account when judging the effects of the other variables. We call these variables **covariables** (or, alternatively, **covariates**). A typical example is an experimental design where samples are grouped into logical or physical blocks. The values of response variables (e.g. species composition) for a group of samples might be similar due to their spatial proximity, so we need to model this influence and account for it in our data. The differences in response variables that are due to the membership of samples in different blocks must be removed (i.e. 'partialled-out') from the model.

But, in fact, almost any explanatory variable can take the role of a covariable. For example, in a project where the effect of management type on butterfly community composition is studied, we might have the localities at different altitudes. The altitude might have an important influence on the butterfly communities, but in this situation we are primarily interested in the management effects. If we remove the effect of the altitude, we might get a clearer picture of the influence that the management regime has on the butterfly populations.

1.6.   Handling missing values in data

Whatever precautions we take, we are often not able to collect all the data values we need: a soil sample sent to a regional lab gets lost, we forget to fill in a particular slot in our data collection sheet, etc.

Most often, we cannot go back and fill in the empty slots, usually because the subjects we study change in time. We can attempt to leave those slots empty, but this is often not the best decision. For example, when recording sparse community data (we might have a pool of, say, 300 species, but the average number of species per sample is much lower), we interpret the empty cells in a spreadsheet as absences, i.e. zero values. But the absence of a species is very different from the situation where we simply forgot to look for this species! Some statistical programs provide a notion of missing values (it might be represented as a word 'NA', for example), but this is only a notational convenience. The actual statistical method must deal further with the fact that there are missing values in the data. Here are few options we might consider:

1. We can remove the samples in which the missing values occur. This works well if the missing values are concentrated in a few samples. If we have,

for example, a data set with 30 variables and 500 samples and there are 20 missing values from only three samples, it might be wise to remove these three samples from our data before the analysis. This strategy is often used by general statistical packages and it is usually called 'case-wise deletion'.

2. On the other hand, if the missing values are concentrated in a few variables that are not deemed critical, we might remove the variables from our data set. Such a situation often occurs when we are dealing with data representing chemical analyses. If 'every thinkable' cation concentration was measured, there is usually a strong correlation among them. For example, if we know the values of cadmium concentration in air deposits, we can usually predict the concentration of mercury with reasonable precision (although this depends on the type of pollution source). Strong correlation between these two characteristics implies that we can make good predictions with only one of these variables. So, if we have a lot of missing values in cadmium concentrations, it might be best to drop this variable from our data.

3. The two methods of handling missing values described above might seem rather crude, because we lose so much of our data that we often collected at considerable expense. Indeed, there are various **imputation** methods. The simplest one is to take the average value of the variable (calculated, of course, only from the samples where the value is not missing) and replace the missing values with it. Another, more sophisticated one, is to build a (multiple) regression model, using the samples with no missing values, to predict the missing value of a variable for samples where the values of the other variables (predictors in the regression model) are not missing. This way, we might fill in all the holes in our data table, without deleting any samples or variables. Yet, we are deceiving ourselves – we only duplicate the information we have. The degrees of freedom we lost initially cannot be recovered.

If we then use such supplemented data in a statistical test, this test makes an erroneous assumption about the number of degrees of freedom (number of independent observations in our data) that support the conclusion made. Therefore, the significance level estimates are not quite correct (they are 'over-optimistic'). We can alleviate this problem partially by decreasing the statistical weight for the samples where missing values were estimated using one or another method. The calculation can be quite simple: in a data set with 20 variables, a sample with missing values replaced for five variables gets a weight 0.75 $(=1.00 - 5/20)$. Nevertheless, this solution is not perfect. If we work only with a subset of the variables (for example, during a stepwise selection of explanatory variables), the samples with any variable being imputed carry the penalty even if the imputed variables are not used.

The methods of handling missing data values are treated in detail in a book by Little & Rubin (1987).

1.7.   Importing data from spreadsheets – WCanoImp program

The preparation of input data for multivariate analyses has always been the biggest obstacle to their effective use. In the older versions of the CANOCO program, one had to understand the overly complicated and un-forgiving format of the data files, which was based on the requirements of the FORTRAN programming language used to create the CANOCO program. Version 4 of CANOCO alleviates this problem by two alternative means. First, there is now a simple format with minimum requirements for the file con-tents (the free format). Second, and probably more important, is the new, easy method of transforming data stored in spreadsheets into CANOCO format files. In this section, we will demonstrate how to use the WCanoImp program for this purpose.

Let us start with the data in your spreadsheet program. While the majority of users will work with Microsoft Excel, the described procedure is applicable to any other spreadsheet program running under Microsoft Windows. If the data are stored in a relational database (Oracle, FoxBASE, Access, etc.) you can use the facilities of your spreadsheet program to first import the data into it. In the spreadsheet, you must arrange your data into a rectangular structure, as laid out by the spreadsheet grid. In the default layout, the individual samples correspond to the rows while the individual spreadsheet columns represent the variables. In addition, you have a simple heading for both rows and columns: the first row (except the empty upper left corner cell) contains the names of vari-ables, while the first column contains the names of the individual samples. Use of heading(s) is optional, because WCanoImp is able to generate simple names there. When using the heading row and/or column, you must observe the lim-itations imposed by the CANOCO program. The names cannot have more than eight characters and the character set is somewhat limited: the safest strategy is to use only the basic English letters, digits, dot, hyphen and space. Neverthe-less, WCanoImp replaces any prohibited characters by a dot and also shortens any names longer than the eight characters. Uniqueness (and interpretability) of the names can be lost in such a case, so it is better to take this limitation into account when initially creating the names.

The remaining cells of the spreadsheet must only be numbers (whole or decimal) or they must be empty. No coding using other kinds of characters is

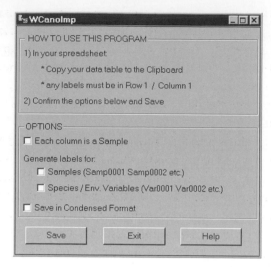

Figure 1-2. The main window of the WCanoImp program.

allowed. Qualitative variables ('factors') must be coded for the CANOCO program using a set of 'dummy variables' – see Section 1.5 for more details.

When the data matrix is ready in the spreadsheet program, you must select the rectangular region (e.g. using the mouse pointer) and copy its contents to the Windows Clipboard. WCanoImp takes the data from the Clipboard, determines their properties (range of values, number of decimal digits, etc.) and allows you to create a new data file containing these values, and conforming to one of two possible CANOCO data file formats. Hopefully it is clear that the requirements concerning the format of the data in a spreadsheet program apply only to the rectangle being copied to the Clipboard. Outside of it, you can place whatever values, graphs or objects you like.

The WCanoImp program is accessible from the Canoco for Windows program menu (**Start** > **Programs** > [*Canoco for Windows folder*]). This import utility has an easy user interface represented chiefly by one dialog box, displayed in Figure 1-2.

The upper part of the dialog box contains a short version of the instructions provided here. Once data are on the Clipboard, check the WCanoImp options that are appropriate for your situation. The first option (*Each column is a Sample*) applies only if you have your matrix transposed with respect to the form described above. This might be useful if you do not have many samples (because Microsoft Excel, for example, limits the number of columns to 256) but a high number of variables. If you do not have names of samples in the first column, you must check the second checkbox (i.e. ask to *Generate labels for: . . . Samples*), similarly check the third checkbox if the first row in the selected spreadsheet rectangle corresponds to the values in the first sample, not to the names of the variables. The last checkbox (*Save in Condensed Format*) governs the actual format used when creating the data file. The default format (used if this option is not checked) is the so-called full format; the alternative format is the condensed format. Unless you are worried about using too much hard disc space, it does not matter what you select here (the results of the statistical methods will be identical, whatever format is chosen).

After you have made sure the selected options are correct, you can proceed by clicking the *Save* button. You must first specify the name of the file to be generated and the place (disc letter and folder) where it will be stored. WCanoImp then requests a simple description (one line of ASCII text) for the data set being generated. This one line then appears in the analysis output and reminds you of the kind of data being used. A default text is suggested in case you do not care about this feature. WCanoImp then writes the file and informs you about its successful creation with another dialog box.

1.8.   Transformation of species data

As will be shown in Chapter 3, ordination methods find the axes representing regression predictors that are optimal for predicting the values of the response variables, i.e. the values in the species data. Therefore, the problem of selecting a transformation for the response variables is rather similar to the problem one would have to solve if using any of the species as a single response variable in the (multiple) regression method. The one additional restriction is the need to specify an identical data transformation for all the response variables ('species'), because such variables are often measured on the same scale. In the unimodal (weighted averaging) ordination methods (see Section 3.2), the data values cannot be negative and this imposes a further restriction on the outcome of any potential transformation.

This restriction is particularly important in the case of the log transformation. The logarithm of 1.0 is zero and logarithms of values between 0 and 1 are negative. Therefore, CANOCO provides a flexible log-transformation formula:

$$y' = \log(A \cdot y + C)$$

You should specify the values of A and C so that after the transformation is applied to your data values (y), the result (y') is always greater or equal to zero. The default values of both A and C are 1.0, which neatly map the zero values again to zero, and other values are positive. Nevertheless, if your original values are small (say, in the range 0.0 to 0.1), the shift caused by adding the relatively large value of 1.0 dominates the resulting structure of the data matrix. You can adjust the transformation in this case by increasing the value of A to 10.0. But the default log transformation (i.e. $\log(y + 1)$) works well with the percentage data on the 0 to 100 scale, or with the ordinary counts of objects.

The question of when to apply a log transformation and when to use the original scale is not an easy one to answer and there are almost as many answers as there are statisticians. We advise you not to think so much about distributional properties, at least not in the sense of comparing frequency histograms of the variables with the 'ideal' Gaussian (Normal) distribution. Rather try to work out whether to stay on the original scale or to log-transform by using the semantics of the hypothesis you are trying to address.

As stated above, ordination methods can be viewed as an extension of multiple regression methods, so this approach will be explained in the simpler regression context. You might try to predict the abundance of a particular species in samples based on the values of one or more predictors (environmental variables, or ordination axes in the context of ordination methods). One can formulate the question addressed by such a regression model (assuming just a single predictor variable for simplicity) as 'How does the average value of species Y change with a change in the environmental variable X by one unit?' If neither the response variable nor the predictors are log-transformed, your answer can take the form 'The value of species Y increases by B if the value of environmental variable X increases by one measurement unit'. Of course, B is then the regression coefficient of the linear model equation $Y = B_0 + B \cdot X + E$. But in other cases, you might prefer to see the answer in a different form, 'If the value of environmental variable X increases by one unit, the average abundance of the species increases by 10%'. Alternatively, you can say 'The abundance increases 1.10 times'. Here you are thinking on a multiplicative scale, which is not the scale assumed by the linear regression model. In such a situation, you should **log-transform the response** variable.

Similarly, if the effect of a predictor (environmental) variable changes in a multiplicative way, **the predictor** variable **should be log-transformed**.

Plant community composition data are often collected on a semi-quantitative estimation scale and the Braun–Blanquet scale with seven levels $(r, +, 1, 2, 3, 4, 5)$ is a typical example. Such a scale is often quantified in the spreadsheets using corresponding ordinal levels (from 1 to 7 in this case). Note that this coding already implies a log-like transformation because the actual cover/abundance differences between the successive levels are generally increasing. An alternative approach to using such estimates in data analysis is to replace them by the assumed centres of the corresponding range of percentage cover. But doing so, you find a problem with the r and + levels because these are based more on the abundance (number of individuals) of the species than on their estimated cover. Nevertheless, using very rough replacements such as 0.1 for r and 0.5 for + rarely harms the analysis (compared to the alternative solutions).

Another useful transformation available in CANOCO is the square-root transformation. This might be the best transformation to apply to count data (number of specimens of individual species collected in a soil trap, number of individuals of various ant species passing over a marked 'count line', etc.), but the log-transformation also handles well such data.

The console version of CANOCO 4.x also provides the rather general 'linear piecewise transformation' which allows you to approximate the more complicated transformation functions using a poly-line with defined coordinates of the 'knots'. This general transformation is not present in the Windows version of CANOCO, however.

Additionally, if you need any kind of transformation that is not provided by the CANOCO software, you might do it in your spreadsheet software and export the transformed data into CANOCO format. This is particularly useful in cases where your 'species data' do not describe community composition but something like chemical and physical soil properties. In such a case, the variables have different units of measurement and different transformations might be appropriate for different variables.

1.9.   Transformation of explanatory variables

Because the explanatory variables ('environmental variables' and 'covariables' in CANOCO terminology) are assumed not to have a uniform scale, you need to select an appropriate transformation (including the popular 'no transformation' choice) individually for each such variable. CANOCO does not provide this feature; therefore, any transformations on the explanatory variables must be done before the data are exported into a CANOCO-compatible data file.

But you should be aware that after CANOCO reads in the environmental variables and/or covariables, it centres and standardizes them all, to bring their means to zero and their variances to one (this procedure is often called *standardization to unit variance*).

2

Experimental design

Multivariate methods are no longer restricted to the exploration of data and to the generation of new hypotheses. In particular, constrained ordination is a powerful tool for analysing data from manipulative experiments. In this chapter, we review the basic types of experimental design, with an emphasis on manipulative field experiments. Generally, we expect that the aim of the experiment is to compare the response of studied objects (e.g. an ecological community) to several treatments (treatment levels). Note that one of the treatment levels is usually a control treatment (although in real ecological studies, it might be difficult to decide what is the control; for example, when we compare several types of grassland management, which of the management types is the control one?). Detailed treatment of the topics handled in this chapter can be found for example in Underwood (1997).

If the response is univariate (e.g. number of species, total biomass), then the most common analytical tools are ANOVA, general linear models (which include both ANOVA, linear regression and their combinations), or generalized linear models. Generalized linear models are an extension of general linear models for the cases where the distribution of the response variable cannot be approximated by the normal distribution.

2.1. Completely randomized design

The simplest design is the completely randomized one (Figure 2-1). We first select the plots, and then randomly assign treatment levels to individual plots. This design is correct, but not always the best, as it does not control for environmental heterogeneity. This heterogeneity is always present as an unexplained variability. If the heterogeneity is large, use of this design might decrease the power of the tests.

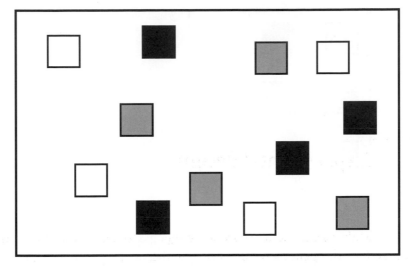

Figure 2-1. Completely randomized design, with three treatment levels and four replicates (or replications).

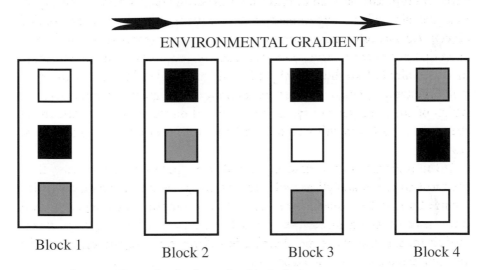

Figure 2-2. The randomized complete blocks design.

2.2. Randomized complete blocks

There are several ways to control for environmental heterogeneity. Probably the most popular one in ecology is the randomized complete blocks design. Here, we first select the blocks so that they are internally as homogeneous as possible (e.g. rectangles with the longer side perpendicular to the environmental gradient, Figure 2-2). The number of blocks is equal to the number

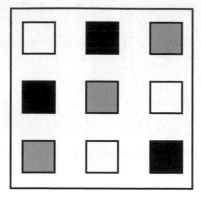

Figure 2-3. Latin square design.

of replications. Each block contains just one plot for each treatment level, and their spatial position within a block is randomized.

If there are differences among the blocks,* this design provides a more powerful test than the completely randomized design. On the other hand, when applied in situations where there are no differences among blocks, the power of the test will be lower in comparison to the completely randomized design, because the number of degrees of freedom is reduced. This is particularly true for designs with a low number of replications and/or a low number of levels of the experimental treatment. There is no consensus among statisticians about when the block structure can be ignored if it appears that it does not explain anything.

2.3.   Latin square design

Latin square design (see Figure 2-3) assumes that there are gradients, both in the direction of the rows and the columns of a square. The square is constructed in such a way that each column and each row contains just one of the levels of the treatment. Consequently, the number of replications is equal to the number of treatments. This might be an unwanted restriction. However, more than one Latin square can be used. Latin squares are more popular in agricultural research than in ecology. As with randomized complete blocks, the environmental variability is powerfully controlled. However, when there is no such variability (i.e. that explainable by the columns and rows), then the test is weaker than a test for a completely randomized design, because of the loss of degrees of freedom.

---

* And often there are: the spatial proximity alone usually implies that the plots within a block are more similar to each other than to the plots from different blocks.

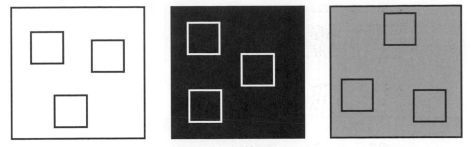

Figure 2-4. Pseudoreplications: the plots are not replicated, but within each plot several subsamples are taken.

2.4.   Most frequent errors – pseudoreplications

Pseudoreplications (Figure 2-4) are among the most frequent errors in ecological research (Hurlbert 1984). A possible test, performed on the data collected using such a design, evaluates differences among plot means, not the differences among treatments. In most cases, it can be reasonably expected that the means of contiguous plots are different: just a proximity of subplots within a plot compared with distances between subplots from different main plots suggests that there are some differences between plots, regardless of the treatment. Consequently, the significant result of the statistical test does not prove (in a statistical sense) that the treatment has any effect on the measured response.

2.5.   Combining more than one factor

Often, we want to test several factors (treatments) in a single experiment. For example, in ecological research, we might want to test the effects of fertilization and mowing.

Factorial designs

The most common way to combine two factors is through factorial design. This means that each level of one factor is combined with each level of the second factor. If we consider fertilization (with two levels, fertilized and non-fertilized) and mowing (mown and non-mown), we get four possible combinations. Those four combinations have to be distributed in space either randomly, or they can be arranged in randomized complete blocks or in a Latin square design.

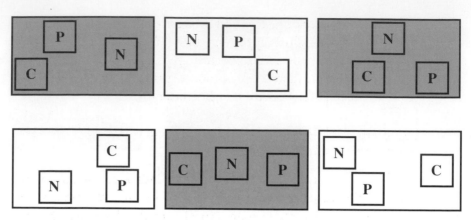

Figure 2-5. The split-plot design. In this example, the effect of fertilization was studied on six plots, three of them on limestone (shaded) and three of them on granite (empty). The following treatments were established in each plot: control (C), fertilized by nitrogen (N), and fertilized by phosphorus (P).

Hierarchical designs

In hierarchical designs, each main plot contains several subplots. For example, we can study the effect of fertilization on soil organisms. For practical reasons (edge effect), the plots should have, say, a minimum size of 5 m × 5 m. This clearly limits the number of replications, given the space available for the experiment. Nevertheless, the soil organisms are sampled using soil cores of diameter 3 cm. Common sense suggests that more than one core can be taken from each of the basic plots. This is correct. However, the individual cores are not independent observations. Here we have one more level of variability – the plots (in ANOVA terminology, the plot is a random factor).

The plots are said to be nested within the treatment levels (sometimes, instead of hierarchical designs, they are called **nested design**). In such a design, a treatment effect is generally tested against the variability on the nearest lower hierarchical level. In the example above, the effect of fertilization is tested against the variability among plots (within a fertilization treatment level), **not** against the variability among soil cores. By taking more soil cores from a plot we do not increase the error degrees of freedom; however, we decrease the variability among the plots within the treatment level, and increase the power of the test in this way.

Sometimes, this design is also called a **split-plot design**. More often, the split-plot design is reserved for another hierarchical design, where two (or more) experimental factors are combined as shown in Figure 2-5.

In this case, we have the **main plots** (with three replications for each level) and the **split-plots**. The main plots are alternatively called whole plots.

The effect of the bedrock must be tested against the variability among the main plots within the bedrock type. Note that this design is different and consequently has to be analysed differently from the factorial design with two factors, bedrock and fertilization – there is one additional level of variability, that of the plots, which is nested within the bedrock type.

2.6. Following the development of objects in time – repeated observations

The biological/ecological objects (whether individual organisms or communities studied as permanent plots) develop in time. We are usually interested not only in their static state, but also in the dynamics of their development, and we can investigate the effect of experimental manipulations on their dynamics. In all cases, the inference is much stronger if the **baseline data** (the data recorded on the experimental objects before the manipulation was imposed) are available. In this case, we can apply a BACI (before after control impact) design (Green 1979).

In designed manipulative experiments, the objects should be part of a correct statistical design; we then speak about replicated BACI designs.

As has been pointed out by Hurlbert (1984), the non-replicated BACI is not a statistically correct design, but it is often the best possibility in environmental impact studies where the 'experiment' is not designed in order to enable testing of the effect. An example is given in Figure 2-6. In this case, we want to assess the effect of a newly built factory on the quality of water in a river (and, consequently, we do not have replicated sites). We can reasonably expect that upstream of the factory, the water should not be affected, so we need a control site above the factory and an impact site below the factory (nevertheless, there might be differences in water quality along the river course). We also need to know the state before the factory starts operating and after the start (but we should be aware that there might be temporal changes independent of the factory's presence). We then consider the changes that happen on the impact site but not on the control site* to be a proof of the factory's impact.

Nevertheless, the observations within a single cell (e.g. on the control site before the impact) are just pseudoreplications. Even so, the test performed in this way is much more reliable than a simple demonstration of either temporal changes on the impact site itself or the differences between the impact and control sites in the time after the impact was imposed.†

---

* In ANOVA terminology, this is the interaction between time and the site factors.
† The only possibility of having a replicated BACI design here would be to build several factories of the same kind on several rivers, which is clearly not a workable suggestion!

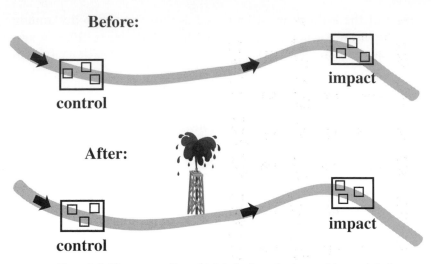

Figure 2-6. The non-replicated BACI (before after control impact) design.

Even if we miss the 'before' situation, we can usually demonstrate a lot by a combination of common sense and some design. For example, in our example case, we can have a series of sites along the river and an abrupt change just between the two sites closest to the factory (one above and one below it), with the differences between the other sites being gradual.* This is often a suggestive demonstration of the factory effect (see Reckhow 1990 for additional discussion).

Another possibility is to repeat the sampling over time and use time points as (possibly correlated) replication (Stewart-Oaten et al. 1986; Ter Braak & Šmilauer 2002).

In designed experiments, the replicated BACI is probably the best solution. In this design, the individual sampling units are arranged in some correct design (e.g. completely randomized or in randomized complete blocks). The first measurement is done before the experimental treatment is imposed (the baseline measurement), and then the development (dynamics) of the objects is followed. In such a design, there should be no differences among the experimental groups in the baseline measurement. Again, the interaction of treatment and time is of greatest interest, and usually provides a stronger test than a comparison of the state at the end of experiment. Nevertheless, even here we have various possible analyses, with differing emphases on the various aspects that we are interested in (see Lindsey 1993). As we control the environmental

* Or similarly, time series from the impacted site, with an abrupt change temporarily coincident with the impact-causing event.

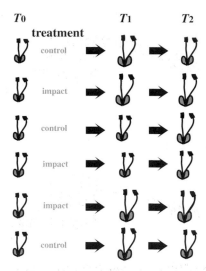

Figure 2-7. Replicated BACI design. The objects (individual plants) are completely randomized and measured at the first time T_0. Thereafter, the impact (fertilization) is imposed and the development is followed (with two sampling times T_1 and T_2 in our example). The null hypothesis we are interested in is: *The growth is the same in the control and impact groups.*

heterogeneity by knowing the initial state, the block structure is usually not as efficient as in the static comparisons. An example is given in Figure 2-7.

One should be aware that, with 5% probability, there will be significant differences between treatments in the baseline data (when tested at $\alpha = 0.05$). It is therefore advisable to test the baseline data before imposing the treatments, and when significant differences are found re-assign the treatments to the units (with the same random process as used originally).

With the continuing development of statistical methods, the designs are becoming more and more flexible. Often the design of our experiment is a compromise between the statistical requirements and the feasibility of the implied manipulations in the field. The use of pseudoreplication in designed experiments is, nevertheless, one of the most common mistakes and can lead to serious errors in our conclusions. Each researcher is advised to have a clear idea about how the data will be analysed before performing the experiment. This will prevent some disappointment later on.

2.7.   Experimental and observational data

Manipulative experiments in ecology are limited both in space and time (Diamond 1986). For larger spatial or temporal scales we have to rely on observational data. Also, each manipulation has inevitably some side-effects and, consequently, the combination of observational data and manipulative experiments is necessary. Generally, the observational data are more often used in hypotheses generation, whereas the data from manipulative experiments are used for hypotheses testing. However, this need not always be so. In manipulations on the community level, we usually affect a whole chain of causally interconnected variables (e.g. in the fertilization experiment in Case Study 3 (Chapter 13), fertilization affected the cover of the crop, which, together with the direct effect of the nutrients, affected the weed community). In replicated manipulative experiments on a community level in space and time, there are many factors that we are not able to control, but still affect the results (weather in a year, bed rock of a site). When the response is species community composition, we often know the biological traits of some species (which we are not able to manipulate). We primarily test the hypothesis concerning change of species composition, but the results very often suggest new hypotheses on the relationship between species traits and species response to the manipulation. All those situations call for careful exploration of data, even in the analysis of manipulative experiments (Hallgren et al., 1999).

Also in non-manipulative studies, we often collect data to test a specific hypothesis. It is useful to collect data in a design that resembles the manipulative experiment. Diamond (1986) even coined the term 'natural experiment', for the situations in nature that resemble experimental designs and can be analysed accordingly (for example, to compare the abundance of spiders on islands with and without lizards). According to Diamond, such situations are under-utilized: 'While field experimentalists are laboriously manipulating species abundances on tiny plots, analogous ongoing manipulations on a gigantic scale are receiving little attention (e.g. expansion of parasitic cowbirds in North America and the West Indies, decimation of native fish by introduced piscivores, and on-going elimination of American elms by disease).' In non-manipulative situations, however, one can never exclude the possibility of confounding effects – consequently, special attention should be paid to sampling design that minimizes this danger.

3
Basics of gradient analysis

The methods for analysing species composition are usually divided into gradient analysis and classification. The term **gradient analysis** is used here in the broad sense, for any method attempting to relate species composition to the (measured or hypothetical) environmental gradients.

Traditionally, the classification methods, when used in plant community ecology, were connected with the **discontinuum approach** (or vegetation unit approach) or sometimes even with the Clementsian superorganismal approach, whereas the methods of gradient analysis were connected with the **continuum concept** or with the Gleasonian individualistic concept of (plant) communities (Whittaker 1975). While this might reflect the history of the methods, this distinction is no longer valid. The methods are complementary and their choice depends mainly on the purpose of a study. For example, in vegetation mapping some classification is necessary. Even if there are no distinct boundaries between adjacent vegetation types, we have to cut the continuum and create distinct vegetation units for mapping purposes. Ordination methods can help find repeatable vegetation patterns and discontinuities in species composition, and show any transitional types, etc. These methods are now accepted even in phytosociology. Also, the methods are no longer restricted to plant community ecology. They became widespread in most studies of ecological communities with major emphasis on species composition and its relationship with the underlying factors. In fact, it seems to us that the advanced applications of gradient analysis are nowadays found outside the vegetation sciences, for example in hydrobiological studies (see the bibliographies by Birks et al. 1996, 1998).

3.1.   Techniques of gradient analysis

Table 3-1 provides an overview of the problems solved by gradient analysis and related methods (the methods are categorized according to Ter Braak & Prentice 1988). The methods are selected mainly according to the data that are available for the analysis and according to the desired result (which is determined by the question we are asking).

The goal of **regression** is to find the dependence of a univariate response (usually the quantity of a species, or some synthetic characteristics of a community, such as diversity or biomass) on environmental (explanatory) variables. As the environmental variables can also be categorical ones, this group of methods also includes analysis of variance (ANOVA). By **calibration** we understand the estimation of values of environmental characteristics based on the species composition of a community. Typical examples are the estimates based on Ellenberg indicator values ('Zeigerwerte', Ellenberg 1991), or estimates of water acidity based on the species composition of diatom communities (Batterbee 1984; Birks 1995). To use calibration procedures, we need to know a priori the species' responses to the environmental gradients being estimated.

The goal of **ordination** is to find axes of the greatest variability in the community composition (the ordination axes) for a set of samples and to visualize (using an ordination diagram) the similarity structure for the samples and species. It is often expected that the ordination axes will coincide with some measurable environmental variables, and when those variables are measured we usually correlate them with the ordination axes. The aim of **constrained ordination** is to find the variability in species composition that can be explained by the measured environmental variables. Ordination and constrained ordination are also available in **partial** versions (not shown in Table 3-1), as **partial ordination** and **partial constrained ordination.** In partial analyses, we first subtract the variability in the species composition explainable by the **covariables** and then perform a (constrained) ordination on the residual variability. In **hybrid ordination** analyses, first x (x is usually 1, 2 or 3) ordination axes are constrained and the remaining axes are unconstrained.

The environmental variables and the covariables can be both quantitative and categorical.

3.2.   Models of species response to environmental gradients

The species response to a continuous environmental variable can be described by a rich variety of shapes. Modern regression methods deal

Table 3-1. *Methods of gradient analysis*

Data used in calculations		A priori knowledge of species–environment relationships	Method	Result
No of envir. variables	No of species			
1, n	1	No	Regression	Dependence of the species on environmental variables
None	n	Yes	Calibration	Estimates of environmental values
None	n	No	Ordination	Axes of variability in species composition
1, n	n	No	Constrained ordination	Variability in species composition explained by the environmental variables Relationship of environmental variables to the species axes

Methods categorized according to Ter Braak & Prentice (1986).

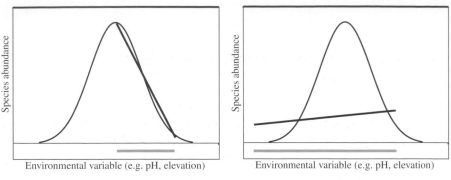

Figure 3-1. Comparison of the suitability of linear approximation of a unimodal response curve over a short part of the gradient (left diagram) and over a larger part of the gradient (right diagram). The horizontal bar shows the part of the gradient we are interested in.

with this variety and also handle response variables characterized by various distributions (see Chapter 8). However, the regression can also be a part of the algorithm of multivariate analyses, where many regression models are fitted simultaneously. For this purpose, we need a relatively simple model, which can be easily fitted to the data. Two models of species response to environmental gradient are frequently used: the model of a linear response and that of a unimodal response. The linear response is the simplest approximation, whereas the unimodal response model assumes that the species has an optimum on the environmental gradient. To enable simple estimation, the unimodal model assumes that the relationship is symmetrical around the species optimum (see Hutchinson 1957, for the concept of resource gradients and species optima). When using ordination methods, we must first decide which of the two models should be used. Generally, both models are just approximations, so our decision depends on which of the two approximations is better for our data. Even the unimodal response is a simplification: in reality (Whittaker 1967), the response is seldom symmetrical, and also more complicated response shapes can be found (e.g. bimodal ones). Moreover, the method of fitting the unimodal model imposes further restrictions.

Over a short gradient, a linear approximation of any function (including the unimodal one) works well, but over a long gradient the approximation by the linear function is poor (Figure 3-1). Even if we have no measured environmental variables, we can expect that, for relatively homogeneous data, the underlying gradient is short and the linear approximation appropriate.

3.3.   Estimating species optima by the weighted averaging method

The linear response is usually fitted by the classical method of (least squares) regression. For the unimodal response model, the simplest way to estimate the species optimum is by calculating the weighted average (WA(Sp)) of the values of environmental variables in the n samples where the species is present. The species importance values (abundances) are used as the weights in calculating the average:

$$\text{WA(Sp)} = \frac{\sum\limits_{i=1}^{n} \text{Env}_i \times \text{Abund}_i}{\sum\limits_{i=1}^{n} \text{Abund}_i}$$

where Env_i is the value of environmental variable in the ith sample, and Abund_i is the abundance of the species in the ith sample.[*] If needed, the species tolerance (the width of the bell-shaped curve) can be calculated as the square root of the weighted mean of the squared differences between the species optimum and the actual values in a sample. The value is analogous to standard deviation (and is a basis for the definition of SD units for measuring the length of ordination axes, see Section 10.2):

$$\text{SD} = \sqrt{\frac{\sum\limits_{i=1}^{n} (\text{Env}_i - \text{WA(Sp)})^2 \times \text{Abund}_i}{\sum\limits_{i=1}^{n} \text{Abund}_i}}$$

The method of the weighted averaging is reasonably good when the whole range of a species' distribution is covered by the samples. Consider the dependence displayed in Figure 3-2. If the complete range is covered, then the estimate is correct (Table 3-2).

On the other hand, if only a part of the range is covered, the estimate is biased. The estimate is shifted in the direction of the tail that is not truncated. An example is presented in Table 3-3.

---

[*] Another possibility is to explicitly state the functional form of the unimodal curve and estimate its parameters by the methods of non-linear regression, but this option is more complicated and not suitable for the simultaneous calculations that are usually used in ordination methods.

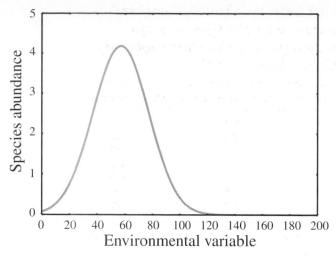

Figure 3-2. Example of a unimodal species response with a range completely covering the response curve.

When the covered portion of the gradient is short, most of the species will have their distributions truncated, and the optimum estimates will be biased. The longer the axis, the more the species will have their optima estimated correctly. We can reasonably expect that the more homogeneous the data, the shorter the gradient.

> The techniques based on the linear response model are suitable for homogeneous data sets; the weighted averaging techniques are suitable for more heterogeneous data.

The decision is usually made on the basis of gradient length in detrended correspondence analysis (DCA), which estimates the heterogeneity in community composition (see Section 4.3).

Table 3-2. *Estimation of species optimum from the response curve displayed in Figure 3-2, using a weighted averaging algorithm when a complete range of species distribution is covered*

Environmental value	Species abundance	Product
0	0.1	0
20	0.5	10
40	2.0	80
60	4.2	252
80	2.0	160
100	0.5	50
120	0.1	12
Total	9.4	564

$$\text{WA(Sp)} = \frac{\sum_{i=1}^{n} \text{Env}_i \times \text{Abund}_i}{\sum_{i=1}^{n} \text{Abund}_i} = 564/9.4 = 60$$

Table 3-3. *Estimation of species optimum from the response curve displayed in Figure 3-2, using a weighted averaging algorithm when only a part of the range of species distribution is covered*

Environmental value	Species abundance	Product
60	4.2	252
80	2.0	160
100	0.5	50
120	0.1	12
Total	6.8	474

$$\text{WA(Sp)} = \frac{\sum_{i=1}^{n} \text{Env}_i \times \text{Abund}_i}{\sum_{i=1}^{n} \text{Abund}_i} = 474/6.8 = 69.7$$

3.4.   Calibration

The goal of calibration is to estimate the values of environmental descriptors on the basis of species composition.* This procedure can be used only if we know beforehand the behaviour of species in relation to the environmental variables to be estimated. Theoretically, the procedure assumes that we have a **training set** with both species and environmental data available, which are used to estimate the relationships of species and environment (e.g. species optima with respect to environmental gradients). Those quantified relationships are then used to estimate the unknown environmental characteristics for samples, where only the species composition is known.

The most commonly used method is **weighted averaging**. For this purpose, the estimates of species optima in relation to selected environmental gradients must be available. These estimated optima values are sometimes called **indicator values**. For example in Central Europe, the indicator values of most species are available (on relative scales) for light, nitrogen availability, soil moisture, etc. (e.g. Ellenberg 1991).[†] The environmental value in a sample (WA(Samp)) can then be estimated as a weighted average of the indicator values of all the s present species, their abundances being used as weights:

$$\text{WA(Samp)} = \frac{\sum_{i=1}^{s} \text{IV}_i \times \text{Abund}_i}{\sum_{i=1}^{s} \text{Abund}_i}$$

where IV_i is the indicator value of ith species (presumably its optimum) and Abund_i is the abundance of ith species in a sample. The procedure is based on the assumption of existence of species optima (and, consequently, of the

---

* A common-sense motivation: the presence of polar bears indicates low temperatures; the presence of *Salicornia* species, high soil salinity; and the presence of common stinging nettle, a high concentration of soil nitrogen.

[†] However, these species indicator values are **not** based on the analyses of training sets, but rather on personal experience of the authors.

Table 3-4. *Estimation of nitrogen availability for two samples, Sample 1 and Sample 2*

	Nitrogen IV	Sample 1	IV × abund.	Sample 2	IV × abund.
Drosera rotundifolia	**1**	2	2	0	0
Andromeda polypofila	**1**	3	3	0	0
Vaccinium oxycoccus	**1**	5	5	0	0
Vaccinium uliginosum	**3**	2	6	1	3
Urtica dioica	**8**	0	0	5	40
Phalaris arundinacea	**7**	0	0	5	35
Total		12	16	11	78
Nitrogen (WA)		1.333		7.090	
		(= 16/12)		(= 78/11)	

The nitrogen indicator values (IV) are listed first (according to Ellenberg, 1991). The a priori known indicator values are shown in bold style (in the Nitrogen IV column), the collected data and the calculations are shown in standard typeface.

unimodal species' responses). However, the procedures for calibration based on the assumption of a linear response also exist (even if they are seldom used in ecology).

An example of the calibration procedure is given in Table 3-4.

Opinion on the use of indicator values and calibration differs among ecologists. It is generally argued that it is more reliable to measure the abiotic environment directly than to use a calibration. However, for historical records (not only relevés (samples) done in the past, but also palynological and paleolimnological evidence, etc.) the calibration provides information that may not be completely reliable, but it is often the only available.

Calibration is also one of the steps used in more complicated analyses, such as ordination methods.

3.5. Ordination

The problem of an unconstrained ordination can be formulated in two ways:

1. Find a configuration of samples in the ordination space so that the distances between samples in this space do correspond best to the dissimilarities of their species composition. This is explicitly done by multidimensional scaling (MDS) methods (Kruskal 1964; Legendre & Legendre 1998). The metric MDS (also called principal coordinates analysis, see Section 6.4) considers the samples to be points in a multidimensional space (e.g. where species are the axes and the

position of each sample is given by the corresponding species abundance). Then the goal of ordination is to find a projection of this multidimensional space into a space with reduced dimensionality that will result in a minimum distortion of the spatial relationships. Note that the result is dependent on how we define the 'minimum distortion'. In non-metric MDS (NMDS), we do not search for projection rules, but using a numerical algorithm we seek for a configuration of sample points that best portrays the rank of inter-sample distances (see Section 6.5).

2. Find 'latent' variable(s) (ordination axes) that represent the best predictors for the values of all the species. This approach requires the model of species response to such latent variables to be explicitly specified: the linear response model is used for linear ordination methods and the unimodal response model for weighted averaging methods. In linear methods, the sample score is a linear combination (weighted sum) of the species scores. In weighted averaging methods, the sample score is a weighted average of the species scores (after some rescaling).

Note: The weighted averaging algorithm contains an implicit standardization by both samples and species. In contrast, we can select in linear ordination the standardized and non-standardized forms.

The two formulations may lead to the same solution.[*] For example, principal component analysis can be formulated either as a projection in Euclidean space, or as a search for latent predictor variables when linear species responses are assumed.

In the CANOCO program, the approach based on formulation 2 is adopted. The principle of ordination methods can be elucidated by their algorithm. We will use the weighted averaging methods as an example. We try to construct the 'latent' variable (the first ordination axis) so that the fit of all the species using this variable as the predictor will be the best possible fit. The result of the ordination will be the values of this latent variable for each sample (called the **sample scores**) and the estimate of species optimum on that variable for each species (the **species scores**). Further, we require that the species optima be correctly estimated from the sample scores (by weighted averaging) and the sample scores be correctly estimated as weighted averages of the species scores (species optima). This can be achieved by the following iterative algorithm:

---

[*] If samples with similar species composition were distant on an ordination axis, such an axis could not serve as a good predictor of their species composition.

- **Step 1** Start with some (arbitrary) initial site scores $\{x_i\}$
- **Step 2** Calculate new species scores $\{y_i\}$ by (weighted averaging) regression from $\{x_i\}$
- **Step 3** Calculate new site scores $\{x_i\}$ by (weighted averaging) calibration from $\{y_i\}$
- **Step 4** Remove the arbitrariness in the scale by standardizing site scores (stretch the axis)
- **Step 5** Stop on convergence, else go to Step 2

We can illustrate the calculation by an example, presented in Table 3-5. The data table (with three samples and four species) is displayed using bold letters. The initial, arbitrarily chosen site scores are displayed in italics. From those, we calculate the first set of species scores by weighted averaging, SPWA1 (Step 2 above). From these species scores, we calculate new sample scores (SAWA1), again by weighted averaging (Step 3). You can see that the axis is shorter (with the range from 1.095 to 8.063 instead of from 0 to 10). The arbitrariness in the scale has to be removed by linear rescaling and the rescaled sample scores SAWA1resc are calculated (Step 4 above):

$$x_{rescaled} = \frac{x - x_{min}}{x_{max} - x_{min}} \times \text{length}$$

where x_{max} and x_{min} are the maximum and minimum values of x in the non-rescaled data and 'length' is the desired length of the ordination axis. Note that the length parameter value is arbitrary, 10.0 in our case. This is true for some ordinations but there are methods where the length of the axis reflects the heterogeneity of the data set (see e.g. detrended correspondence analysis (DCA) with Hill's scaling, Section 10.2). Now, we compare the original values with the newly calculated set. Because the values are very different, we continue by calculating new species scores, SPWA2. We repeat the cycle until the consecutive sample scores SAWANresc and SAWAN+1resc have nearly the same value (some criterion of convergence is used, for example we considered in our case values 0.240 and 0.239 to be almost identical).

Table 3-5 was copied from the file *ordin.xls*. By changing the initial values in this file, you can confirm that the final values are completely independent of the initial values. In this way, we get the first ordination axis.[*] The higher ordination axes are derived in a similar way, with the additional constraint that they have to be linearly independent of all the previously derived axes.

---

[*] In the linear methods, the algorithm is similar, except the regression and calibration are not performed using weighted averaging, but in the way corresponding to the linear model – using the least squares algorithm.

Table 3-5. *Calculation of the first ordination axis by the weighted averaging (WA) method. Further explanation is in the text*

	Samp1	Samp2	Samp3	SPWA1	SPWA2	SPWA3	SPWA4
Cirsium	0	0	3	10.000	10.000	10.000	10.000
Glechoma	5	2	1	2.250	1.355	1.312	1.310
Rubus	6	2	0	1.000	0.105	0.062	0.060
Urtica	8	1	0	0.444	0.047	0.028	0.027
Initial value	*0*	*4*	*10*				
SAWA1	1.095	1.389	8.063				
SAWA1resc.	0.000	0.422	10.000				
SAWA2	0.410	0.594	7.839				
SAWA2resc.	0.000	0.248	10.000				
SAWA3	0.376	0.555	7.828				
SAWA3resc.	0.000	0.240	10.000				
SAWA4	0.375	0.553	7.827				
SAWA4resc.	0.000	0.239	10.000				

The whole method can be also formulated in terms of matrix algebra and eigenvalue analysis. For practical needs, we should note that the better the species are fitted by the ordination axis (the more variability the axis explains), the less the axis 'shrinks' in the course of the iteration procedure (i.e. the smaller is the difference between SAWA and SAWAresc). Consequently, the value

$$\lambda = \frac{x_{max} - x_{min}}{length}$$

is a measure of the explanatory power of the axis and, according to the matrix algebraic formulation of the problem, it is called the **eigenvalue.** Using the method described above, each axis is constructed so that it explains as much variability as possible, under the constraint of being independent of the previous axes. Consequently, the eigenvalues decrease with the order of the axis.[*]

3.6.   Constrained ordination

Constrained ordination can be best explained within the framework of ordinations defined as a search for the best explanatory variables (i.e. problem formulation 2 in the previous section). Whereas in unconstrained ordinations we search for any variable that best explains the species composition (and this

---

[*] A strictly mathematically correct statement is that they do not increase, but for typical data sets they always do decrease.

Table 3-6. *Basic types of ordination techniques*

	Linear methods	Weighted averaging
Unconstrained	Principal components analysis (PCA)	Correspondence analysis (CA)
Constrained	Redundancy analysis (RDA)	Canonical correspondence analysis (CCA)

variable is taken as the ordination axis), in constrained ordinations the ordination axes are weighted sums of environmental variables. Numerically, this is achieved by a slight modification of the above algorithm, in which we add one extra step:

- **Step 3a** Calculate a multiple regression of the site scores $\{x_i\}$ on the environmental variables and take the fitted values of this regression as new site scores

Note that the fitted values in a multiple regression are a linear combination of the predictors and, consequently, the new site scores are linear combinations of the environmental variables. The fewer environmental variables we have, the stricter is the constraint. If the number of environmental variables is greater than the number of samples minus 2, then the ordination is unconstrained.

The unconstrained ordination axes correspond to the directions of the greatest variability within the data set. The constrained ordination axes correspond to the directions of the greatest data set variability that can be explained by the environmental variables. The number of constrained axes cannot be greater than the number of environmental variables.

3.7.   Basic ordination techniques

Four basic ordination techniques can be distinguished based on the underlying species response model and whether the ordination is constrained or unconstrained (see Table 3-6, based on Ter Braak & Prentice, 1988). The unconstrained ordination is also called *indirect gradient analysis* and the constrained ordination is called *direct gradient analysis*.

For weighted averaging methods, detrended versions exist (i.e. detrended correspondence analysis (DCA) implemented in the legendary DECORANA program, Hill & Gauch 1980, and detrended canonical correspondence analysis (DCCA, see Section 4.5). For all methods, partial analyses exist. In partial analyses, the effect of covariables is first removed and the analysis is then performed on the remaining variability. The number of constrained axes cannot exceed the number of environmental variables. When we use just one environmental variable, only the first ordination axis is constrained and the remaining axes are unconstrained.

The **hybrid analyses** represent a 'hybrid' between constrained and unconstrained ordination methods. In standard constrained ordinations, there are as many constrained axes as there are independent explanatory variables and only the additional ordination axes are unconstrained. In a hybrid analysis, only a pre-specified number of canonical axes are calculated and any additional ordination axes are unconstrained. In this way, we can specify the dimensionality of the solution of the constrained ordination model.

With 'release' from the constraint (after all the constrained axes are calculated), the procedure is able to find a 'latent' variable that may explain more variability than the previous, constrained ones. Consequently, in most cases, the first unconstrained axis explains more then the previous constrained axis and, therefore, the corresponding eigenvalue is higher than the previous one.

3.8.   Ordination diagrams

The results of an ordination are usually displayed as ordination diagrams. Plots (samples) are displayed by points (symbols) in all the methods. Species are shown by arrows in linear methods (the direction in which the species abundance increases) and by points (symbols) in weighted averaging methods (estimates of the species optima). Quantitative environmental variables are shown by arrows (direction in which the value of environmental variable increases). For qualitative environmental variables, the centroids are shown for individual categories (the centroid of the plots, where the category is present). More information on the interpretation of ordination diagrams can be found in Chapter 10.

Typical examples of ordination diagrams produced from the results of the four basic types of ordination methods are shown in Figure 3-3.

3.9.   Two approaches

Having both environmental data and data on species composition, we can first calculate an unconstrained ordination and then calculate a regression of the ordination axes on the measured environmental variables (i.e. to project the environmental variables into the ordination diagram) or we can calculate directly a constrained ordination. **The two approaches are complementary and both should be used!** By calculating the unconstrained ordination first,

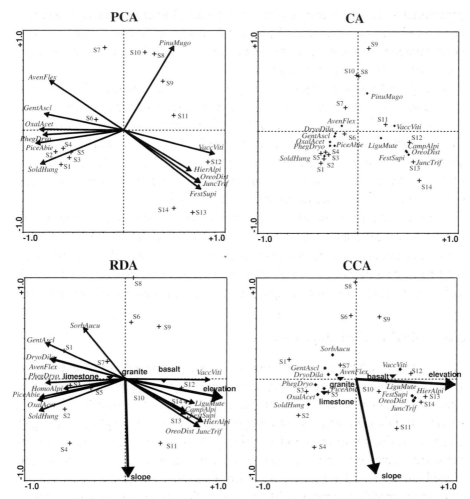

Figure 3-3. Examples of typical ordination diagrams. Analyses of species composition of 14 samples (S1 to S14), located on an elevation gradient, in plots differing in slope and located on various bedrock types. The species are labelled by the first four letters of the generic name and the first four letters of the specific name. The quantitative environmental variables are elevation and slope, the qualitative variable is the bedrock type, represented by three categories (basalt, granite, limestone). Samples (relevés) are displayed as crosses. Partly artificial data.

we do not miss the main part of the variability in species composition, but we can miss that part of the variability that is related to the measured environmental variables. By calculating a constrained ordination, we do not miss the main part of the biological variability explained by the environmental variables, but we can miss the main part of the variability that is not related to the measured environmental variables.

When you publish your results, always be careful to specify the method of analysis used. From an ordination diagram, it is impossible to distinguish between constrained and unconstrained ordinations; because many authors do not follow the convention of using arrows for species in linear methods, even distinction between linear and unimodal methods is not unequivocal.

3.10. Testing significance of the relation with environmental variables

In an ordinary statistical test, the value of the test statistic calculated from the data is compared with the expected distribution of the statistic under the null hypothesis being tested. Based on this comparison, we estimate the probability of obtaining results as different (or even more different) from those expected under the null hypotheses, as in our data. The distribution of the test statistic is derived from the assumption about the distribution of the original data (this is why we assume the normal distribution of the response residuals in least squares regression). In CANOCO, the distribution of the test statistic* under the null hypothesis of independence is not known. This distribution depends on the number of environmental variables, their correlation structure, the distribution of the species abundances, etc. However, the distribution can be simulated and this is done in a **Monte Carlo permutation test.**

In this test, an estimate of the distribution of the test statistic under the null hypothesis is obtained in the following way. The null hypothesis states that the response (the species composition) is independent of the environmental variables. If this hypothesis is true, then it does not matter which observation of explanatory variable values is assigned to which observation of species composition. Consequently, the values of the environmental variables are randomly assigned to the individual samples of species composition, ordination analysis is done with this permuted ('shuffled') data set, and the value of the test statistic is calculated. In this way, both the distribution of the response variables and the correlation structure of the explanatory variables remain the same in the real data and in the null hypothesis simulated data. The significance level (probability of type I error) of this test is then calculated as:

$$P = \frac{n_x + 1}{N + 1}$$

---

* The F-ratio, used in the later versions of CANOCO, is a multivariate counterpart of the ordinary F-ratio, the eigenvalue was used in previous versions.

Table 3-7. *Example of permutation test for a simple linear regression*

Plant height	Nitrogen (as measured)	1st permutation	2nd permutation	3rd permutation	4th permutation	5th etc.
5	3	3	8	5	5	...
7	5	8	5	5	8	...
6	5	4	4	3	4	...
10	8	5	3	8	5	...
3	4	5	5	4	3	...
F-value	10.058	0.214	1.428	4.494	0.826	0.###

where n_x is the number of permutations where the test statistic was not lower in the random permutation than in the analysis of original data, and N is the total number of permutations. This test is completely distribution-free. This means that it does not depend on any assumption about the distribution of the species abundance values. More thorough treatment of permutation tests in general can be found in Legendre & Legendre (1998), pp. 20–26.

The permutation scheme can be 'customized' according to the experimental design from which the analysed data sets come. The above description corresponds to the basic version of the Monte Carlo permutation test, but more sophisticated approaches are actually used in the CANOCO program – see Chapter 5 and the Canoco for Windows manual (Ter Braak & Šmilauer, 2002).

3.11. Monte Carlo permutation tests for the significance of regression

The permutation tests can be used to test virtually any relationship. To illustrate its logic, we will show its use for testing the significance of a simple regression model that describes the dependence of plant height on the soil nitrogen concentration.

We know the heights of five plants and the content of nitrogen in the soil in which they were grown (see Table 3-7). We calculate the regression of plant height on nitrogen content. The relationship is characterized by the F-value of the analysis of variance of the regression model. Under some assumptions (normality of the data), we know the distribution of the F-values under the null hypothesis of independence (F-distribution with 1 and 4 degrees of freedom). Let us assume that we are not able to use this distribution (e.g. normality is violated). We can simulate this distribution by randomly assigning the nitrogen values to the plant heights. We construct many random permutations and for each one we calculate the regression (and corresponding F-value) of the plant height on the (randomly assigned values of) nitrogen content. As the nitrogen values were assigned randomly to the plant heights, the distribution of the F-values corresponds to the null hypothesis of independence. Significance of the regression is then estimated as:

$$\frac{1 + \text{no. of permutations where} \, (F \geq 10.058)}{1 + \text{total number of permutations}}$$

The F-ratio in CANOCO has a similar meaning as the F-value in ANOVA of the regression model and the Monte Carlo permutation test is used in an analogous way.

Using the Canoco for Windows 4.5 package

4.1.   Overview of the package

The Canoco for Windows package is composed of several programs. In this section we summarize the role of each program in data analysis and in the interpretation of results. The following sections then deal with some typical aspects of this software use. This chapter is not meant to replace the documentation distributed with the Canoco for Windows package, but provides a starting point for efficient use of this software.

Canoco for Windows 4.5

This is the central piece of the package. Here you specify the data you want to use, the ordination model to apply, and the analysis options. You can also select subsets of the explained and explanatory variables to use in the analysis or change the weights for the individual samples. All these choices are collected in a CANOCO project.

Canoco for Windows allows one to use quite a wide range of ordination methods. The central ones are the linear methods (principal components analysis, PCA, and redundancy analysis, RDA) and the unimodal methods (correspondence analysis, CA, detrended correspondence analysis, DCA, and canonical correspondence analysis, CCA), but based on them you can use CANOCO to apply other methods such as multiple discriminant analysis (canonical variate analysis, CVA) or **metric** multidimensional scaling (principal coordinates analysis, PCoA) to your data set. From the widely used ordination methods, only non-metric multidimensional scaling (NMDS) is missing.

CANOCO 4.5

This program can be used as a less user-friendly, but slightly more powerful alternative to the Canoco for Windows program. It represents a non-graphical console (with the text-only interface) version of this software. The user interface is similar to that of previous versions of the CANOCO program (namely versions 3.x), but the functionality of the original program has been extended.

The console version is much less interactive than the Windows version. If a mistake is made and an incorrect option specified, there is no way to change incorrect entries. The program must be terminated and the analysis restarted.

Nevertheless, there are a few 'extras' in the console version's functionality. The most important is the acceptance of 'irregular' design specifications. You can have, for example, data repeatedly collected from permanent plots in three localities. If the duration of data collection varies among sites, there is no way to specify this design to the Windows version of the package so as to ensure correct permutation restrictions during the Monte Carlo permutation test. The console version allows you to specify the arrangement of the samples (in terms of spatial and temporal structure and/or the general split-plot design) independently for each block of samples.

Another advantage of the console version is its ability to read the analysis specification, which is normally entered by the user as the answers to individual program questions, from a 'batch' file. Therefore, it is possible to generate such batch files and run few to many analyses at the same time. This option is obviously an advantage only for experienced users.

WCanoImp and CanoImp.exe

The functionality of the WCanoImp program is described in Section 1.7. When using this program, you are limited by the memory capacity of the Windows' Clipboard and also by the capacity of the sheet of your spreadsheet program. For Microsoft Excel, you cannot have more than 255 columns of data (assuming the first column contains the row labels); therefore, you must limit yourself to a maximum of 255 variables or 255 samples. The other dimension is more forgiving: Microsoft Excel 97 allows 65,536 rows.

If your data do not fit into those limits, you can either split the table, export the individual parts using WCanoImp and then merge the resulting CANOCO files using the CanoMerge program (see next subsection, CanoMerge) or you can use the console form (command line) of the WCanoImp program – program file **canoimp.exe**. Both WCanoImp and canoimp programs have the same purpose and the same functionality, but there are two important differences.

The first difference is that the input data must be stored in a text file for canoimp. The content of this file should be in the same format as data pasted from a spreadsheet program; namely, a textual representation of the spreadsheet cells, with transitions between columns marked by TAB characters and the transition between rows marked by new-line characters. The simplest way to produce such an input file for the canoimp.exe program is to proceed as if using the WCanoImp program, up to the point that data are copied to the Clipboard. From there, switch to the WordPad program (in Windows 9x or Windows ME) or the Notepad program (in Windows NT, Windows 2000, or Windows XP), create a new document, and select the *Edit > Paste* command. Then save the document as an ASCII file (WordPad program supports other formats as well). Alternatively, save the sheet from the spreadsheet program using the *File > Save As . . .* command and select the format called *Text file (Tab separated)* (the format name can somewhat vary across the Excel versions). Note that this works flawlessly only if the data table is the only content of the spreadsheet document.

The second difference between the WCanoImp utility and the canoimp program is the way in which options are selected. Whilst both programs have the same options available, in WCanoImp they are easily selected by ticking checkboxes; however, in canoimp the options must be typed on a command line together with the names of the input and output files. So, a typical execution of the program from the command prompt looks like this:

```
d:\canoco\canoimp.exe -C -P inputdta.txt output.dta
```

where the –C option means the output is in condensed format, while the –P option means a transposition of the input data (i.e. the rows in the input text file represent variables). The TAB-separated format will be read from inputdta.txt and CanoImp will create a new data file (and overwrite any existing file with the same name) named output.dta, using the CANOCO condensed format.

CanoMerge

Program CanoMerge can be used for three different purposes, often combined at the same time.

1. The primary task of this program is to merge two or more datafiles, containing the same set of samples but different variables. This is useful in situations described in the previous subsection, where we want to start with a data table stored in a Microsoft Excel spreadsheet and both the number of samples and the number of variables (e.g. species) exceed 255. You can divide such a table into several sheets, each with the same number of rows (representing the samples), but with the set of variables

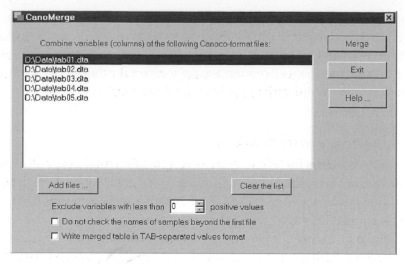

Figure 4-1. CanoMerge program user interface.

split over the sheets (e.g. the first 255 species in the first sheet, the next 255 species in the second one, etc.). If you export such partial tables separately from the Excel file into the data files with CANOCO format, you can use the CanoMerge application (see Figure 4-1) to join them into one large datafile, which will contain all the species.

You select all the files containing the partial tables, using the *Add files* button and the program creates the new merged data table after you clicked the *Merge* button.

2. The CanoMerge program can be used to export your datafiles into ASCII files with the TAB-separated file format. This is useful particularly if you do not have your data available in the Excel spreadsheets or if the data table is too large to fit there. The TAB-separated format files are accepted by most statistical packages. Note that you can transform even a single input datafile in this way: you simply specify just the one file in the dialog shown after the *Add files* button was clicked.

3. Finally, you can create a new datafile from an existing one(s), which will contain only the frequently occurring species. CanoMerge allows you to filter the set of input variables (species, presumably, in this context) that are copied to the output file depending on the number of non-zero values (occurrences) they have in the data. This is quite a useful feature, because in the CANOCO program you can delete particular species or make them supplementary, but there is no easy way to exclude less frequent species, beyond the 'downweighting of rare species' feature. You can exclude species with fewer than X occurrences using the *Exclude variables with less than* X *positive values* field in the CanoMerge program.

CanoDraw 3.1

The CanoDraw 3.1 program was distributed with the Canoco for Windows package when it first appeared, in version 4.0. The new versions of the CANOCO package (from version 4.5) contain the CanoDraw for Windows program and we will stick to the new CanoDraw version in this book.

CanoDraw for Windows

CanoDraw for Windows is the program for visualizing ordination results obtained with the CANOCO program and exploring them in the context of the original data used in the ordination analyses and any additional information available for the project samples or variables.

Similar to Canoco for Windows, your work with CanoDraw is also centred on projects. Each CanoDraw project is usually based on one CANOCO analysis. To create it, you specify the name of an existing (and already analysed) CANOCO project. CanoDraw reads the analysis options from the *.con* (project) file, imports the analysis results (from the CANOCO *.sol* file), and also imports the original data files referred by the CANOCO project (files with species data, and with environmental and supplementary variables, and with covariables, if available). After the CanoDraw project has been created and saved into a file (using the name with a *.cdw* extension), it is independent of the existence and location of the original files that were used or produced by CANOCO.

CanoDraw for Windows supports the creation of ordination diagrams, as well as of various XY and contour plots, in which fitted regression models can be displayed.

PrCoord

The PrCoord program calculates the full solution of principal coordinates analysis (for which the abbreviation PCoA is used in this book, but be aware that elsewhere the abbreviation PCO may be used), also known as metric multidimensional scaling. Principal coordinates analysis allows you to summarize (dis-)similarity (or distance) among a set of samples in a few dimensions, starting from an $N \times N$ matrix of dissimilarity coefficients (or distance measures). In this respect, PCoA is a kind of unconstrained ordination, like PCA or CA, but with it you can specify the distance measure used by the method, while in PCA and CA the type of measure is implied (Euclidean distance in PCA and chi-square distance in CA). The CANOCO program alone provides the possibility to calculate PCoA on an input matrix of intersample distances, but you must have the matrix available in the CANOCO-compatible format and it is also quite difficult to obtain all the principal coordinates in CANOCO (by default, only the first four axes are calculated).

More advanced methods starting from the PCoA solution, such as distance-based RDA (Legendre & Anderson 1999), require the use of all the principal coordinates, and PrCoord program stores them all* in a datafile in CANOCO-compatible format. In addition, not only can PrCoord read an existing matrix of intersample distances, it can also calculate it for a relatively wide range of distance measures, from a primary data matrix submitted in a file with CANOCO format.

4.2.   Typical flow-chart of data analysis with Canoco for Windows

Figure 4-2 shows a typical sequence of actions taken when analysing multivariate data. You first start with the data sets recorded in a spreadsheet and export them into CANOCO-compatible data files, using the WCanoImp program. In the Canoco for Windows program, you either create a new CANOCO project or clone an existing one using the *File > Save as...* command. Cloning retains all the project settings and we can change only the settings that need to be changed.

The decision about the ordination model, shown in Figure 4-2, refers to asking new questions about your data, as well as to iterative improvement of the ordination model that is used to address those various questions. The consequent changes in ordination models then involve different standardization of response variables, change in the set of used environmental variables and co-variables, optional transformations of predictor variables, etc.

Each CANOCO project is represented by two windows (views). The **Project View** summarizes the most important project properties (e.g. type of ordination method, dimensions of the data tables and names of the files in which the data are stored). Additionally, the Project View features a column of buttons providing shortcuts to the commands most often used when working with the projects: running analysis, modifying project options, starting the CanoDraw program, saving the analysis log, etc. The **Log View** records the users' actions on the project and the output created during the project analysis. Some of the statistical results provided by CANOCO are available only from this log. Other results are stored in the 'SOL file', containing the actual ordination scores. You

---

* More precisely, PrCoord stores all the principal coordinates with positive eigenvalues after an eventual adjustment for negative eigenvalues, as described by Legendre & Anderson, 1999.

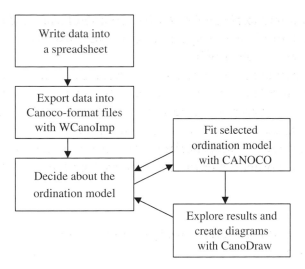

Figure 4-2. A simplified flow-chart of data analysis with Canoco for Windows.

can extend the contents of the Log View by entering new text (comments) into the log: the Log View works as a simple text editor.

You can define the project settings using the Project setup wizard. This wizard can be invoked by clicking the *Options* button in Project View. CANOCO displays the first page from a sequence of pages containing various pieces of information the program needs to know to run the analysis. This sequence is not a static one; the page displayed at a certain time depends on the choices made in the preceding pages. For example, some of the options are specific to linear ordination methods, so these pages are displayed only if a linear method (PCA or RDA) was chosen. You proceed between pages using the *Next* button. You may return to the preceding pages using the *Back* button. Some of the critical choices to be made with the setup wizard are discussed in more detail later in this chapter. On the last page, the *Finish* button replaces the *Next* button. After you click this button, the changes in the options are applied to the project. If you were defining a new project, CANOCO now asks for the name of the file in which the project will be saved.

Once the project is defined, the analysis can be performed (the data analysed) by clicking the *Analyze* button in Project View (or, alternatively, by using the shortcut button from the toolbar or using the menu command). If successful, the results are stored in the solution file (its name was specified on the second Project setup wizard page) and additional information is placed into the Log View, where it might be inspected. In the Log View, you can find a statistical summary for the first four ordination axes, information on the correlation between the environmental variables and the ordination axes, an indication of any outlier observations and the results of the Monte Carlo permutation tests. Part of this information is essential for performing certain tasks, but nothing needs to be retained for plotting the ordination diagrams with CanoDraw. CanoDraw needs only the results stored in the solution file.*

With CanoDraw, you can explore the ordination results and combine them with the information from the original data. You can specify not only what is plotted, but also various plot properties – range of axes, which items are plotted, contents of the attribute plots, etc. The resulting diagrams can be further adjusted: you can change the type of symbols, their size and colours, fonts used for individual labels, line types, etc.

4.3.   Deciding on the ordination method: unimodal or linear?

This section provides a simple-to-use 'cookbook' for deciding whether you should use ordination methods based on a model of linear species response to the underlying environmental gradient or weighted-averaging (WA) ordination methods, corresponding to a model of unimodal species response. The presented recipe is unavoidably over-simplified, so it should not be followed blindly.

In the Canoco for Windows project that you use to decide between the unimodal and linear methods, you try to match as many choices that you will make in the final analysis as possible. If you have covariables, you use them here as well; if you use only a subset of the environmental variables, use the subset here too. If you log-transform (or square-root-transform) species data, you do it here as well.

For this trial project, you select the weighted-averaging method with detrending. This means either DCA for indirect gradient analysis or detrended canonical correspondence analysis (DCCA) for a constrained analysis. Then you select the option 'detrending by segments' (which also implies the Hill's scaling of ordination scores) and the other options as in the final analysis and run the analysis. You then look at the analysis results stored in Log View. At the end of the log, there is the Summary table and in it is a row starting with 'Lengths of gradient', looking like the following example:

```
Lengths of gradient : 2.990 1.324 .812 .681
```

---

* There is an exception to this rule: when you are creating scores for principal response curves (PRC, see Section 9.3) with CanoDraw, you must have the Log View contents available in a text file. Use the *Save log* button to create such a file.

The gradient length measures the beta diversity in community composition (the extent of species turnover) along the individual independent gradients (ordination axes). Now you locate the largest value (the longest gradient) and if that value is larger than 4.0, you should use unimodal methods (DCA, CA, or CCA). Use of a linear method would not be appropriate, since the data are too heterogeneous and too many species deviate from the assumed model of linear response (see also Section 3.2). On the other hand, if the longest gradient is shorter than 3.0, the linear method is probably a better choice (not necessarily, see Section 3.4 of Ter Braak & Šmilauer 2002). In the range between 3 and 4, both types of ordination methods work reasonably well.

> When deciding whether to use the linear or unimodal type of ordination method, you must take into account another important difference among them. The unimodal methods always implicitly work with standardized data. CCA, CA or DCA methods summarize variation in the relative frequencies of the response variables (species). An important implication of this fact is that these methods cannot work with 'empty' samples, i.e. records in which no species is present. Also, the unimodal methods cannot be used when the response variables do not have the same units.

4.4.   PCA or RDA ordination: centring and standardizing

Centring and standardization choices (see Section 6.2) made in linear ordination methods in the CANOCO program may substantially change the outcome of the analyses. In fact, you can address different questions about your data (see the discussion in Section 14.3). Here we provide brief guidance for selecting the right centring and standardization options.

This project setup wizard page is displayed for linear ordination methods (PCA and RDA) and refers to the manipulations with the species data matrix before the ordination is calculated.

Centring by samples (left column of the wizard page, Figure 4-3) results in a zero average for each row. Similarly, centring by species (right column of the wizard page, Figure 4-3) results in a zero average for each column. Centring by species is obligatory for the constrained linear method (RDA) or for any **partial** linear ordination method (i.e. where covariables are used).

Standardization by species (samples) results in the norm of each column (each row) being equal to one. The **norm** is the square root of sum of squares of the column (row) values. If you apply both centring and standardization, the centring is done first. After centring **and** standardizing by species, the columns

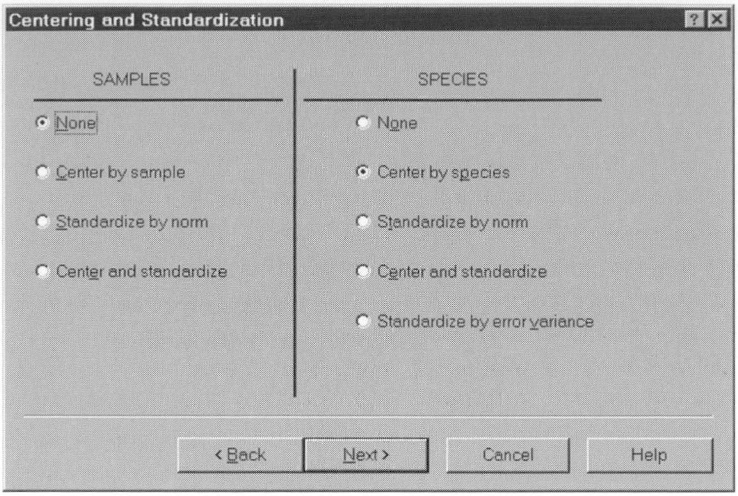

Figure 4-3. Centring and standardization options in the Project setup wizard.

represent variables with zero average and unit variance. As a consequence, PCA performed on the species data then corresponds to a 'PCA on a matrix of correlations' (between species). If you do not standardize by species norm, the resulting PCA is the 'PCA on variance-covariance matrix'. Note that the standardization by species in situations where species differ substantially in their average frequency and/or average quantity in the data can put rather too much weight on rare species in the analysis.

Also note that you must always standardize by species (and probably not standardize or centre by samples) in situations, where your response variables ('species data') differ in their measurement scales; for example, when you study the variability of various physico-chemical parameters with PCA.

If you have environmental variables available in the ordination method (always in RDA and optionally in PCA), you can select the standardization by **error variance.** In this case, CANOCO calculates, separately for each species, how much of its variance was not explained by the environmental variables (and covariables, if there are any). The inverse of that *error variance* is then used as the species weight. Therefore, the better a species is described by the environmental variables provided, the greater weight it has in the final analysis.

4.5.   DCA ordination: detrending

The detrending of second and higher ordination axes of correspondence analysis (leading to detrended correspondence analysis, DCA) is often

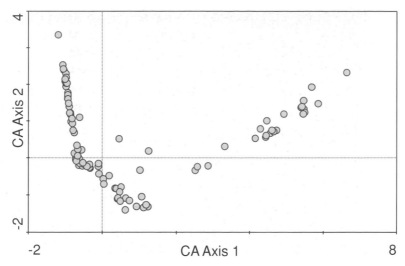

Figure 4-4. Scatter of samples from a correspondence analysis, featuring the so-called arch effect. You can see the 'arch' by rotating your book by 180 degrees.

used to cope with the so-called arch effect, illustrated in Figure 4-4 by a diagram with sample positions on the first two axes of correspondence analysis (CA).

The positions of the samples on the second (vertical) axis are strongly (but not linearly) dependent on their positions on the first (horizontal) axis. This effect can be interpreted as a limitation of the method in which the consecutive axes are made mutually independent (only the linear independence is sought) or, alternatively, as a consequence of the projection of the non-linear relations of response variables to the underlying gradients into a linear Euclidean drawing space (see Legendre & Legendre 1998, pp. 465–472 for more detailed discussion). **Detrending by segments** (Hill & Gauch 1980), while lacking a convincing theoretical basis and considered as inappropriate by some authors (e.g. Knox 1989 or Wartenberg et al. 1987), is the most often used approach for making the recovered compositional gradient straight (linear). When you opt for detrending in your unimodal ordination model, CANOCO displays the *Detrending Method* page, illustrated in Figure 4-5.

Use of detrending by segments is not recommended for unimodal ordination methods where either covariables or environmental variables are present. In such cases if a detrending procedure is needed, **detrending by polynomials** is the recommended choice. You are advised to check the Canoco for Windows manual for more details on deciding among the polynomials of second, third, or fourth degree (Section 7.6 of Ter Braak & Šmilauer 2002).

The detrending procedure is usually not needed for a constrained unimodal ordination. If an arch effect occurs in CCA, this is usually a sign of some

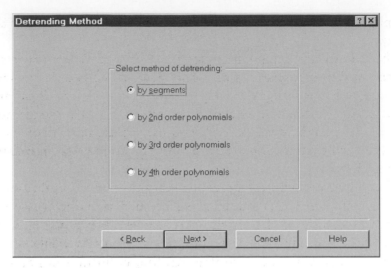

Figure 4-5. Detrending method selection in the Project setup wizard.

redundant environmental variables being present. There may be two or more environmental variables strongly correlated (positively or negatively) with each other. If you retain only one variable from such a group, the arch effect often disappears. The selection of a subset of environmental variables with a low cross-correlation can be performed using the forward selection of the environmental variables in Canoco for Windows.

4.6.   Scaling of ordination scores

The most important result of an ordination method applied to particular data is the ordination diagram. Using this diagram, it is theoretically possible to reconstruct (with a certain level of error) not only the primary data table (the species data), but also the matrix of (dis-)similarities among our samples and the species correlation matrix.* Nobody attempts to recover such data from an ordination diagram because the measured values are available, but it is useful to go part way when interpreting the content of the ordination diagram and generating interesting research hypotheses. The precision of conclusions about the similarity of samples, relations between species and/or environmental variables, etc. depends partly on the relative scaling of the scores on the individual ordination axes. One kind of scaling is more favourable if the viewer's attention is focused on the relation between

---

* For linear ordination methods; the matrix of dissimilarities (Chi-square distances) among samples and among species is implied by the results of unimodal ordination methods.

samples, whereas another is favoured if the relation between species is to be interpreted.

The alternative scaling options allow you to support alternative aspects of the visual interpretation of the ordination results. The choices made in these pages do not change other aspects of the analysis (such as the amount of variability explained by individual ordination axes, the strength and direction of relations between the explanatory variables and ordination axes, or the type I error estimates provided by the significance tests in constrained ordinations).

The options are somewhat similar in linear and unimodal ordination methods (see Figure 4-6). First, the decision must be made as to whether the viewer's attention will focus on the samples (this includes the comparison of classes of samples, as portrayed by the nominal environmental variables) or on the species during the interpretation of the resulting ordination diagrams.

Then, for the linear model, you must decide whether the lengths of arrows should mirror the differences in the abundances of individual species in the data (with dominant species generally having longer arrows than those with small abundance values) or whether the abundance of individual species should be transformed to a comparable scale. This latter choice corresponds to the *correlation biplots*. If you select for the species scores to be divided by the species standard deviation, then the length of each species arrow expresses how well the values of that species are approximated by the ordination diagram. If you decide, on the other hand, that the scores are **not** to be divided by the standard deviation, the arrow length shows the variability in that species' values across the displayed ordination (sub)space.

In the case of unimodal ordination, you should also decide on the method for interpreting the ordination diagrams. For data with very long composition gradients (with large beta diversity across samples),* the **distance rule** is more appropriate and so is Hill scaling. In other cases, **biplot** scaling provides ordination diagrams that can be interpreted in a more quantitative way.

If you plot separate scatter-plots of samples or species, your choice of focus upon species or samples is irrelevant, because CanoDraw for Windows offers automatic adjustment of CANOCO-produced scores. In this way, the resulting scatter-plots are always scaled in the optimum way. Note, however, that this effect cannot be achieved in ordination diagrams where several types of entities are plotted together.

More information about the scaling of scores and its relation to interpretation of ordination diagrams is provided in Sections 10.1 and 10.2.

---

* Ordination axes with length in SD (species-turnover) units approximately more than 3.

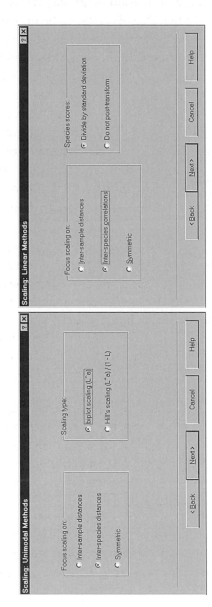

Figure 4-6. Scaling options for linear and for unimodal methods in the Project setup wizard.

4.7.   Running CanoDraw for Windows 4.0

The Canoco for Windows application takes your specification of multi-variate statistical analysis and applies it to the data file(s) you have chosen. In CANOCO, you can do an interactive selection of explanatory variables and review the overall quality of the fitted ordination model (amount of variability explained by particular ordination axes) as well as the results of significance tests (Monte Carlo permutation tests). But the ecologically interpretable results are obtained mostly from the ordination diagrams, created from the output information provided by CANOCO. This is the niche for the CanoDraw program. CanoDraw can be started directly from the Canoco for Windows program, using the *CanoDraw* button displayed in the Project View window.

When creating a new diagram with CanoDraw, you should concentrate on the graph contents first: the range of the diagram axes, the plotted items, the appropriate type of labelling of items in the diagram, etc. First and foremost, you must decide what kind of diagram you want to produce. When summarizing a CANOCO project where both species (community composition) and environmental data were used, the biplot diagram, which includes species and environmental variables, is the most appropriate first choice. Before creating that diagram, two important decisions must be made about its content:

- The qualitative (categorical) environmental variables (i.e. factors) are best represented by the centroid scores for the individual dummy variables. For a factor, each dummy variable corresponds to one factor level. As its name suggests, each centroid symbol represents the centroid (barycentre) of the scores of samples belonging to that particular class (factor level). CanoDraw uses these scores only for the variables that you specified as nominal variables. You can select them (separately for environmental and supplementary variables), using the two menu commands in the *Project > Nominal variables* submenu.
- For typical community composition data, it often does not make sense to display the scores of all species. Some are so rare in the data that no relevant information can be provided about their ecological preferences. Other species might not be characterized well by the explanatory variables used in the analysis. We usually try to define a subset of species appearing in an ordination diagram using a combination of two criteria: species fit and species weight. Both are accessible from the *Inclusion Rules* page of the dialog box invoked by the *Project > Settings* menu command.
  1. **Species fit** represents the percentage of variability in species values explained by the ordination subspace onto which the species scores are projected (usually the first two ordination axes). Informally, the

species fit can be characterized as the quality of the description of the 'behaviour' of species values, derived from the particular combination of ordination axes. This characteristic is not available for the unimodal-model-based analyses where detrending by segments was applied.

2. **Species weight** is normally usable only for unimodal ordination models and is equal to the sum of abundances of the species taken over all the samples.

To limit the set of plotted species, we usually increase the lower values of both criteria ranges for unimodal ordination methods. However, for a linear ordination method where all species have the same weight the only option is to increase the minimum species fit. Note that you can selectively manipulate the selection range concerning the fit of species to horizontal (lower-order) axis of a displayed ordination diagram. This is particularly useful when you use only one explanatory variable in constrained analysis and, therefore, only the first axis reflects the relation of species abundances to such a predictor.

Once the diagram is created (use the *Create > Simple Ordination Plot* command and select the first item on the list for the species-environmental variables bi-plot), you can reposition the individual labels, modify their properties (font type, size, colour) or add some extra information (additional text, arrows pointing to some notable species, etc.). Alternatively, if you are not satisfied with the diagram contents (too many species are displayed, for example), you can adjust the program settings using the commands in the *View* and *Project* sub-menus and recreate the graph using the *Create > Recreate graph* command. Note, however, that in such a case all post-creation changes to your graph will be discarded.

You can save the current state of any CanoDraw graph in a file with a *.cdg* extension. CanoDraw provides basic support for managing multiple graphs for each project and also for working with multiple projects at the same time. CanoDraw graphs can be printed directly from the program, copied to other applications through the Windows Clipboard, or exported in a variety of exchange formats (BMP, PNG, Adobe Illustrator, and Enhanced Metafile Format).

4.8.   New analyses providing new views of our data sets

Working with multivariate data usually has an iterative aspect. You might find it useful to modify the explanatory variables, based on your exploration of the results you get from the initial analyses. This includes a selection of the subset of environmental (explanatory) variables based on the knowledge you gain during the forward selection of environmental variables in CANOCO.

Based on the exploration of the relationship of explanatory variables to the ordination axes and to individual species, you can transform the environmental variables to maximize their linear relation with the gradients being recovered by the ordination model.

In some cases, the ordination results reveal two or more distinct groups of samples, based on their different composition. In that case, it is often useful to enhance the results by further separate analyses for each of those groups. In a direct gradient analysis, this often leads to a different picture: the environmental variables identifiable with the differences **between** the groups are frequently different from the variables acting **within** such groups.

Another type of follow-up analysis is used when you identify some environmental variables that are important for explaining part of the structure in the species data. In such a case you can turn those variables into covariables and either test the additional effects of other potentially interesting explanatory variables (partial direct gradient analysis) or just inspect the 'residual' variation and try to informally interpret any pattern found in that way (partial indirect gradient analysis).

5

Constrained ordination and permutation tests

In this chapter, we discuss constrained ordination and related approaches: stepwise selection of a set of explanatory variables, Monte Carlo permutation tests and variance partitioning procedure.

5.1.   Linear multiple regression model

We start this chapter with a summary of the classical linear regression model, because it is very important for understanding the meaning of almost all aspects of direct gradient analysis.

Figure 5-1 presents a simple linear regression used to model the dependence of the values of a variable Y on the values of a variable X. The figure shows the fitted regression line and also the difference between the true (observed) values of the response variable Y and the fitted values ($\hat{Y}$; on the line). The difference between these two values is called the **regression residual** and is labelled as e in Figure 5-1.

An important feature of all **statistical models** (including regression models) is that they have two main components. The **systematic component** describes the variability in the values of the response variable(s) that can be explained by one or more explanatory variables (predictors), using a parameterized function. The simplest function is a linear combination of the explanatory variables, which is the function used by (general) linear regression models. The **stochastic component** is a term referring to the remaining variability in the values of the response variable, not described by the systematic part of the model. The stochastic component is usually defined using its probabilistic, distributional properties.

We judge the quality of the **fitted** model by the amount of the response variable's variance that can be described by the systematic component. We usually compare this to the amount of the unexplained variability that is represented

60

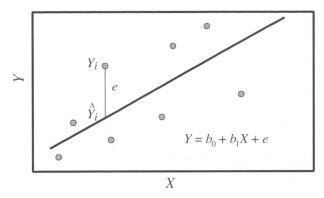

Figure 5-1. Graphical representation of a simple linear regression model.

by the stochastic component. We tend to present such a regression model only where all the predictors significantly contribute to its quality. We can select an appropriate subset of the predictors using a **stepwise selection**. Its most frequently used type is **forward selection**. We start forward selection with a null model with no predictors, assuming that no part of the variability in the response variable can be predicted and that it represents only stochastic variability. We then select a single predictor from the set of available variables – usually the one leading to a regression model that explains the greatest amount of the variability.[+]

Even when adding the best predictor, its contribution to the model might be just due to chance: if we randomly swapped that predictor's values, even such a nonsense variable would still explain a positive amount of the variability in the values of the response variable, so we must test the contribution of the considered candidate predictor. If the contribution of the selected candidate predictor is judged to be non-random ('statistically significant'), we can accept it and repeat the process and try to find another good predictor in the pool of the remaining variables. We should again test its contribution and stop this selection as soon as the 'best' among the remaining predictors is not 'good enough'.

5.2.   **Constrained ordination model**

In Chapter 3, the linear and unimodal methods of **unconstrained ordination** (PCA and CA, respectively) were defined as methods seeking one or more (mutually independent) gradients representing 'optimal' predictors for fitting the regression models of linear or unimodal species response. The

---

[+] Under the constraint of using only one predictor, of course.

optimality is limited by the assumptions of these methods and judged over all the species in the primary data.

The methods of **constrained ordination**[*] have an identical task, but the gradients that these methods 'are allowed to find' are further restricted. Namely, these gradients must be linear combinations of the provided explanatory (environmental) variables. Therefore, we try to explain again the abundances of (all) individual species using synthetic variables (ordination axes), but these variables are further defined using the values of the observed (environmental) characteristics.

In this way, the methods of direct gradient analysis resemble the model of multivariate multiple regression. But in such a regression with m response variables (number of species) and p predictors (the environmental variables), we must estimate $m \cdot p$ parameters (regression coefficients) from the data. This is not the case with a constrained ordination, where the effect of predictors on the response variables is 'focused' through a low number of intermediate gradients – the ordination axes, called here the **canonical** (or constrained) **axes**. There are as many canonical axes as there are **independent** explanatory variables, but we typically use only the first few of them.

In CANOCO, we quite often perform **partial** analyses when we use so-called **covariables** in addition to the environmental (explanatory) variables. These **covariables** represent the effects we want to account for, and therefore remove from the resulting ordination model. Covariables are used in a similar context in analysis of variance. There we often specify quantitative covariables in addition to factors. In traditional regression analysis, the notion of covariables is not frequently used, as the difference among them and the 'true' explanatory variables is only in the way we look at them. Both types are explanatory variables in a regression model (and in an ANOVA or ordination model) and it is only the role we attribute to them that differs.

5.3.   RDA: constrained PCA

We will illustrate the concepts introduced in the previous section by means of redundancy analysis (RDA, van den Wollenberg 1977), a constrained form of the linear ordination method of principal components analysis (PCA, Hotelling 1933). We will use a very simple setup, where we try to determine only the first ordination axis (the first principal component) and we have two environmental variables (z_1 and z_2) available which define ('constrain') the ordination axis of RDA.

---

[*] Also called direct gradient analysis or canonical ordination methods.

Both PCA and RDA methods try to find values of a new variable (we will denote it X) that represents 'the best' predictor for the values of all the species (response variables). The value of that new variable for the ith sample is X_i and we use it to predict the values of the kth species in the ith sample based on the following equation:

$$Y_{ik} = b_{0k} + b_{1k}X_i + e_{ik}$$

Here both PCA and RDA must estimate two sets of parameters: the values X_i, which are the **sample scores** on the first ordination axis, and the regression coefficients for each species (b_{1k}), which are the **species scores** on the first ordination axis. The additional parameter for each species (b_{0k}) represents the 'intercept' of the fitted regression line and we can eliminate its estimation by centring the primary data by species (i.e. subtracting the species averages, see Section 4.4).

Here the similarity between PCA and RDA ends, because in the constrained model of RDA the values of the sample scores are further constrained. They are defined as a linear combination of the explanatory variables. In our example, we have two such variables, z_1 and z_2, so this constraint can be written as:

$$X_i = c_1 z_{i1} + c_2 z_{i2}$$

Note that the parameters estimated here (c_j for jth environmental variable) do not represent the scores of the environmental variables that are usually plotted in the ordination diagrams. Rather, they represent the **Regr** scores in the output from a CANOCO analysis.

We may combine the above two equations into a single one, further illustrating the relation of constrained ordination to multiple multivariate regression:

$$Y_{ik} = b_{0k} + b_{1k}c_1 z_{i1} + b_{1k}c_2 z_{i2} + e_{ik}$$

The expressions $b_{ik} \cdot c_j$ represent the actual coefficients of a multiple multivariate regression model, which describes the dependency of abundance of a particular species k on the values of (two) environmental variables z_1 and z_2. If we fit such models independently for each of m species, we need to estimate $m \cdot p$ regression coefficients (for p environmental variables). But in RDA, the estimated coefficients are constrained by their definition (i.e. they are defined as $b \cdot c$): if we use only the first canonical axis, we need to estimate only $m + p$ parameters (the b_{1k} and c_j coefficients, respectively). To use the first two

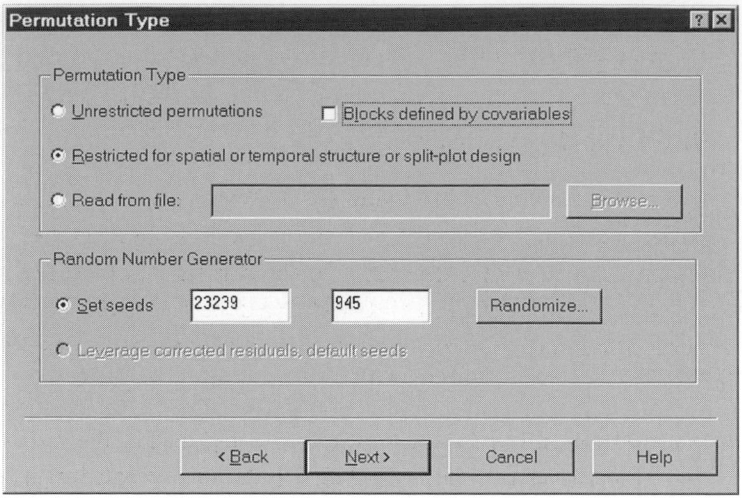

Figure 5-2. Introductory page of the Project setup wizard for selecting the Monte Carlo permutation type.

constrained axes, we need $2 \cdot (m + p)$ parameters.[*] This is, for most datasets, still a large saving compared with the $m \cdot p$ parameters.

5.4.   Monte Carlo permutation test: an introduction

CANOCO has the ability to test the significance of constrained ordination models described in the preceding section, using **Monte Carlo permutation tests.** These statistical tests relate to the general null hypothesis, stating the independence of the primary (species) data on the values of the explanatory variables. The principles of the permutation tests were introduced in Chapter 3, Sections 3.10 and 3.11, in an example of testing a regression model and using the simplest possible type of permutation, i.e. completely random permutation. But CANOCO provides a rich toolbox of specific setups for the tests applied to data sets with a particular spatial, temporal or logical internal structure, which is related to the experimental or sampling design, as can be seen from Figure 5-2.

This figure shows the first page of a sequence of pages available in the Project setup wizard to specify the properties of the permutation test. The following four sections provide more detailed treatment of permutation testing with the

---

[*] This estimate disregards the remaining freedom in b and c values. For example, using $4 \cdot b$ and $c/4$ instead of b and c would yield the same model. A more careful mathematical argument shows that the number of parameters is somewhat different (Robinson 1973; Ter Braak & Šmilauer 2002, p. 56).

CANOCO program. The practical aspects of the testing are further illustrated in most of the case studies in this book.

5.5.   Null hypothesis model

The null model of the independence between the corresponding rows of the species data matrix and of the environmental data matrix (the rows referring to the same sample) is the basis for the permutation test in CANOCO. While the actual algorithm used in CANOCO does not directly employ the approach described here, we use it to better illustrate the meaning of the permutation test.

- We start by randomly re-shuffling (permuting) the samples (rows) in the environmental data table, while keeping the species data intact. Any combination of the species and environmental data obtained in that way is as probable as the 'true' data set, if the null hypothesis is true.
- For each of the data sets permuted in this manner, we calculate the constrained ordination model and express its quality in a way similar to that used when judging the quality of a regression model. In a regression model, we use an *F*-statistic, which is the ratio between the variability of the response variable explained by the regression model (divided by the number of the model parameters) and the residual variability (divided by the number of residual degrees of freedom). In the case of constrained ordination methods, we use similar statistics, described in more detail in the following section.
- We record the value of such a statistic for each permuted (randomized) version of the data set. The distribution of these values defines the distribution of this test statistic under the null model (the histogram in Figure 5-3). If it is highly improbable that the test statistic value obtained from the actual data (with no permutation of the rows of the environmental data table) comes from that distribution (being much larger, i.e. corresponding to an ordination model of higher quality), we reject the null hypothesis. The probability that the 'data-derived' value of the test statistics originates from the calculated null model distribution then represents the probability of a type I error, i.e. the probability of rejecting a correct null hypothesis.

5.6.   Test statistics

The previous section described the general principle of permutation tests and here we discuss the possible choices for the test statistics used in such tests. We mentioned that such a statistic resembles the F-statistic used in the parametric significance test of a regression model. But the choice of definition for this statistic in constrained ordination is difficult, because of the multi-dimensionality of the obtained solution. In general, the variability in species data, described by the environmental variables, is expressed by more than one canonical axis. The relative importance of the canonical axes decreases from the first up to the last canonical axis, but we can rarely ignore all the canonical axes beyond the first one. Therefore, we can either express the accounted variance using all canonical axes or focus on one canonical axis, typically the first one. This corresponds to the two test statistics available in CANOCO version 3.x and 4.x and the two corresponding permutation tests:

- Test of the first canonical axis uses an F-statistic defined in the following way:

$$F_1 = \frac{\lambda_1}{\text{RSS}/(n - p - q)}$$

  The variance explained by the first (canonical) axis is represented by its eigenvalue (λ_1). The residual sum of squares (RSS) term corresponds to the difference between the total variance in the species data and the amount of variability explained by the first canonical axis (and also by the covariables, if any are present in the analysis). The number of covariables is q. The number of independent environmental variables (i.e. the total number of canonical axes) is p. The total number of ordination axes is n.

- Test of the sum of the canonical eigenvalues, in which the overall effect of p explanatory variables, revealed on (up to) p canonical axes, is used:

$$F_{\text{trace}} = \frac{\sum\limits_{i=1}^{p} \lambda_i / p}{\text{RSS}/(n - p - q)}$$

  The RSS term in this formula represents the difference between the total variability in the species data and the sum of eigenvalues of all the canonical ordination axes (adjusted, again, for the variability explained by covariables, if there are any).

As explained in the previous section, the value of one of the two test statistics, calculated from the original data, is compared with the distribution of the statistic under the null model assumption (with the relation between the species data and the environmental variables subjected to permutations). This is illustrated by Figure 5-3.

In this figure, we see a histogram approximating the shape of the distribution of the test statistic. The histogram was constructed from the F values

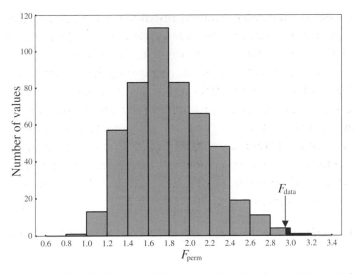

Figure 5-3. The distribution of the F-statistic values from a Monte Carlo permutation test compared wiht the F-statistic value of the 'true' data sets. The black area corresponds to permutation based F-ratio values exceeding the F-ratio value based on the 'true data'.

calculated using the permuted data.[*] The position of the vertical arrow marks the value calculated from the 'real' data. The permutations where the corresponding F-like statistic values are above this level represent the evidence in favour of not rejecting the null hypothesis and their relative frequency represents our estimate of the type I error probability. The actual formula, which was introduced in the Section 3.10, is repeated here:

$$P = \frac{n_x + 1}{N + 1}$$

where n_x is the number of the permutations yielding an F-statistic as large or larger than that from the real data and N is the total number of permutations. The value 1 is added to both numerator and denominator because (under the assumption of the null model) the statistic value calculated from the actually observed data is also considered to come from the null-model distribution and, therefore, to 'vote against' the null hypothesis rejection. The usual choices of the number of permutations (such as 99, 199, or 499) follow from this specific pattern of adding one to both numerator and denominator.

---

[*] The 'true' value was also included, however.

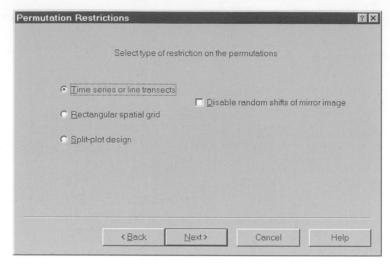

Figure 5-4. Project setup wizard page for selecting restrictions for a permutation test.

5.7.   Spatial and temporal constraints

The way the permutation test was described in the preceding section is correct only when the collected set of samples does not have any implied internal structure; namely, if the samples are sampled randomly, independently of each other. In this case, we can fully randomly re-shuffle the samples, because under the null model each sample's environmental data (values of explanatory variable) can be matched with any other sample's species data with equal probability.

This is no longer true when the 'relatedness' between different samples is not uniform across the whole data set. The basic three families of the internal data set structure recognized by Canoco for Windows are represented by the choices in the setup wizard page illustrated in Figure 5-4.

The samples may be arranged along a linear transect through space or along a time axis. In such cases, the samples cannot be permuted randomly, because we must assume an autocorrelation between the individual observations in both the species and the environmental data. We should not disturb this correlation pattern during the test because our hypothesis concerns the relations between the species and environmental data, not the relations within these data sets. To respect this autocorrelation structure, CANOCO (formally) bends the sequence of samples in both the species and the environmental data into a circular form and a re-assignment is performed by 'rotating' the environmental data band with respect to the species data band. The Canoco for Windows manual should be consulted for more details of this and other permutation restrictions.

A similar spatial autocorrelation occurs when we generalize the location of samples on a linear transect into positioning of samples in space. CANOCO supports only the placement of samples on a rectangular grid.

The most general model of the internal structure for data sets is provided by the **split-plot design**, the last item in the dialog in Figure 5-4. This type of permutation restrictions is described in more detail in the following section and an example of its use is shown in Chapters 15 and 16.

All these restrictions can be further nested within another level representing the blocks. Blocks can be defined in your analysis using nominal covariables and they represent groups of samples that are similar to each other more than to the samples from the other blocks. To account for variation due to blocks, check the *Blocks defined by covariables* option in the dialog illustrated in Figure 5-2. In the permutation test, the samples are permuted only within those blocks, never across the blocks. If we compare a constrained ordination model with a model of the analysis of variance, the blocks can be seen as a random factor with its effect not being interesting for interpretation purposes.

5.8.   Split-plot constraints

The split-plot design restriction available in Canoco for Windows 4.0 and 4.5 allows us to describe a structure with two levels of variability (with two 'error levels') – see Section 2.5.[+]

The higher level of the split-plot design is represented by so-called **whole-plots**. Each of the whole-plots contains the **split-plots**, which represent the lower level of the design (see also Figure 2-5). CANOCO provides great flexibility in permuting samples at the whole-plot and/or the split-plot levels, ranging from no permutation, through spatially or temporally restricted permutations up to free exchangeability at both levels (see the setup wizard page in Figure 5-5). In addition, CANOCO allows for the notion of dependence of the split-plot structure across the individual whole-plots. In this case, the particular permutations of the split-plots within the whole-plots are identical across the whole-plots.

The availability of the permutation types supporting the split-plot design in CANOCO is important because it is also often used for evaluation of data

[+] Another level can be added, in some cases, using permutation within blocks defined by covariables (see Section 5.7).

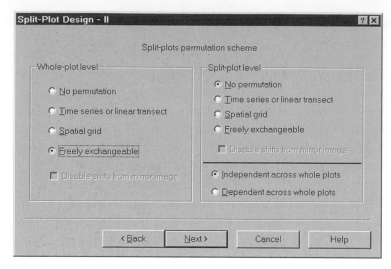

Figure 5-5. Project setup wizard page for specifying permutation restrictions for a split-plot design.

where community composition at sampling plots was repeatedly evaluated at different times. See Chapter 15 for additional details.

5.9.   Stepwise selection of the model

At the end of Section 5.1, the forward selection of explanatory variables for a regression model was described in some detail. The forward selection available in the CANOCO program has the same purpose and methodology. It uses a partial Monte Carlo permutation test to assess the usefulness of each potential predictor (environmental) variable for extending the subset of explanatory variables used in the ordination model.

If we select an interactive ('manual') forward selection procedure in the Canoco for Windows program, CANOCO presents an interactive dialog during the analysis (Figure 5-6).

Figure 5-6 illustrates the state of the forward selection procedure where two best explanatory variables (moisture and manure) were already selected (they are displayed in the lower part of the window). The values in the upper part of the window show that the two selected variables account for approximately 57% (0.624 of 1.098) of the total variability explained by all the environmental variables (i.e. explained when all the variables in the top list are included in the ordination model).

The list of variables in the upper part of the window shows the remaining 'candidate predictors' ordered by the decreasing contribution that the variable

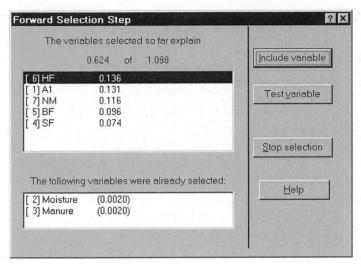

Figure 5-6. Dialog box for the forward selection of environmental variables.

would provide when added to the set of variables already selected. We can see that the variable 'HF' ('hobby farming' type of management) is a hot candidate. It would increase the amount of explained variability from 0.624 to 0.760 (0.624 + 0.136).

To judge whether such an increase is larger than a random contribution, we can use a **partial Monte Carlo permutation test.** In this test, we use the candidate variable as the only explanatory variable (so we get an ordination model with just one canonical axis). The environmental variables already selected (moisture and manure, in our example) are used in this test as covariables, together with any a priori selected covariables. If we reject the null hypothesis for this partial test, we can include that variable in the subset.

The effect of the variable tested in such a context is called its **conditional** (or partial) **effect** and it depends on the variables already selected. But at the start of the forward selection process, when no environmental variable has yet entered the selected subset, we can test each variable separately, to estimate its independent, **marginal effect.** This is the amount of variability in the species data that would be explained by a constrained ordination model using that variable as the only explanatory variable. The discrepancy between the order of variables sorted based on their marginal effects and the order corresponding to a 'blind' forward selection (achieved by always picking the best candidate) is caused by the correlations between the explanatory variables. If the variables were completely linearly independent, both orders would be identical.

If the primary purpose of the forward selection is to find a sufficient subset of the explanatory variables that represents the relation between the species and environmental data, then we have a problem with the 'global' significance level referring to the whole subset selection. If we proceed by selecting the environmental variables as long as the best candidate has a type I error estimate (P) lower than some pre-selected significance level α, then the 'global' type I error probability is, in fact, higher than this level. We do not know how large it is, but we know that it cannot be larger than $N_c \cdot \alpha$. N_c is the maximum number of tests (comparisons) that can be made during the selection process.

The adjustment of the significance threshold levels on each partial test where we select only the variables with a type I error probability estimate less than α/N_c is called a **Bonferroni correction** (Cabin & Mitchell 2000; Rice 1989). Here the value of N_c represents the maximum possible number of steps during the forward selection (i.e. the number of **independent** environmental variables). Use of Bonferroni correction is a controversial issue: some statisticians do not accept the idea of 'pooling' the risk of multiple statistical tests. There are also other methods of correcting the significance levels in partial tests, reviewed in Wright (1992). They are slightly more complicated than Bonferroni correction, but result in more powerful tests.

We can also look at the problem of selecting a subset of 'good' explanatory variables from a larger pool from a different angle: the probability that a variable will be a significant predictor (e.g. of the species composition) purely due to chance equals our pre-specified significance level. If we have, in constrained ordination analysis, 20 environmental variables available, we can expect that on average one of them will be judged significant even if the species are not related to any of them. If we select a subset of the best predictors, drop the other variables, and then test the resulting model, we will very probably get a significant relationship, regardless of whether the environmental variables explain the species composition or not.

Let us illustrate the situation with a simple example. For real data on species composition of plots on an elevation gradient (Lepš et al. 1985; file *tatry.xls*, the *tatry* sheet), we have generated 50 random variables with a uniform distribution of values between zero and a random number between zero and 100 (sheet *tatrand*).[*] Now, we suggest you use those 50 variables as the environmental data. Use CCA (canonical correspondence analysis) with the default settings and the manual forward selection. During the forward selection, use the 'standard' procedure, i.e. test each variable, and finish the forward selection if the best variable has the significance level estimate $P > 0.05$. In this way,

---

[*] The macro used to generate the values of environmental variables is included in the same Excel file so that you can check how the variables were generated.

three variables are selected (var20, var31 and var47). If you now test the resulting model, the global permutation test estimates the type I error probability $P = 0.001$ (with 999 permutations). Clearly, the forward selection (or any stepwise procedure) is a powerful tool for model building. However, when you need to test the significance of the built model, use of an independent data set is necessary. If your data set is large enough, then the best solution is to split the data into a part used for model building, and a part used for model testing (Hallgren et al. 1999).

Another difficulty that we might encounter during the process of forward selection of environmental variables occurs when we have one or more factors coded as dummy variables and used as explanatory (environmental) variables. The forward selection procedure treats each dummy variable as an independent predictor, so we cannot evaluate the contribution of the whole factor at once. This is primarily because the whole factor contributes more than one degree of freedom to the constrained ordination model. Similarly, a factor with K levels contributes $K - 1$ degrees of freedom in a regression model. In a constrained ordination, $K - 1$ canonical axes are needed to represent the contribution of such a factor. While this independent treatment of individual factor levels can make interpretation difficult, it also provides an opportunity to evaluate the extent of differences between the individual classes of samples defined by such a factor. The outcome is partly analogous to the multiple comparison procedure in analysis of variance.[$]

5.10.   Variance partitioning procedure

In the previous section, we explained the difference between the conditional and marginal effects of individual explanatory (environmental) variables upon the species data (the response variables). We also stated that the discrepancy in the importance of explanatory variables, as judged by their marginal effects and their conditional effects, is caused by the correlations between those explanatory variables. Any two explanatory variables that are correlated share part of their effect exercised[*] upon the species data. The amount of the explanatory power shared by a pair of explanatory variables (A and B, say) is equal to the difference between variable A's marginal effect and its conditional effect evaluated in addition to the effect of variable B.[+]

[$] But here we compare the partial effect of a particular factor level with a pooled effect of the not-yet selected factor levels.

[*] In a statistical, not necessarily causal, sense.

[+] The relation is symmetrical – we can label any of the two considered variables as A and the other as B.

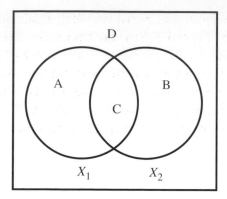

Figure 5-7. Partitioning of the variance in the species data into the contributions of two subsets of environmental variables (A, B, and the shared portion C) and the residual variance (D).

This concept forms the basis of the **variance partitioning**[X] procedure (Borcard et al. 1992). In this procedure, we usually do not quantify the effects of just two explanatory variables: rather, we attempt to quantify the effects (and their overlap) of two or more **groups** of environmental variables representing some distinct, ecologically interpretable phenomena. A separation of the variability of community composition in space and in time might serve as a typical example.

We will describe the variance partitioning procedure using the simplest example with two groups of explanatory variables (X_1 and X_2). Each group contains one or more individual environmental variables. The diagram in Figure 5-7 presents the breakdown of the total variability in the species data according to those two groups of variables.

The area marked as D corresponds to the residual variability, i.e. the variability not explained by the ordination model, which includes the environmental variables from both the X_1 and the X_2 groups. Fraction A represents the partial effect of the variables from group X_1, similarly fraction B represents the partial effect of the variables in group X_2. The effect shared by both groups is the fraction C. The amount of variability explained by group X_1, when ignoring variables from group X_2, is equal to A + C. We estimate the individual fractions A, B and C using partial constrained ordinations.

We estimate A from the analysis where variables from X_1 are used as environmental variables and the variables from X_2 as covariables. Similarly, we estimate B as the sum of the eigenvalues of canonical axes from the analysis where

---

[X] **Variance partitioning** is also called **variation partitioning** (Legendre & Legendre, 1998), which is a more appropriate name if we include unimodal ordination models in our considerations.

X_2 act as environmental variables and X_1 as covariables. We then calculate the size of C by subtracting the sum of A and B from the amount of variability explained by an ordination model with the variables from both X_1 and X_2 acting as explanatory variables. Occasionally, a shared variance fraction (such as C in our example) might have a negative value, indicating that the joint explanatory effect of the two groups of variables is stronger than a sum of their marginal effects. See Legendre and Legendre (1998), p. 533 for a more detailed discussion of this issue.

6

Similarity measures

In many multivariate methods, one of the first steps is to calculate a matrix of similarities (resemblance measures) either between the samples (relevés) or between the species. Although this step is not explicitly done in all the methods, in fact each of the multivariate methods works (even if implicitly) with some resemblance measure. The linear ordination methods can be related to several variants of Euclidean distance, while the weighted-averaging ordination methods can be related to chi-square distances. The resemblance functions are reviewed in many texts (e.g. Gower & Legendre 1986; Legendre & Legendre 1998; Ludwig & Reynolds 1988; Orloci 1978), so here we will introduce only the most important ones.

In this chapter, we will use the following notation: we have n samples (relevés), containing m species. Y_{ik} represents the abundance of the kth species $(k = 1, 2, \ldots, m)$ in the ith sample $(i = 1, 2, \ldots, n)$.

It is technically possible to transpose the data matrix (or exchange i and k in the formulae), thus any of the resemblance functions can be calculated equally well for rows or columns (i.e. for samples or species). Nevertheless, there are generally different kinds of resemblance functions suitable for expressing the (dis)similarity among samples, and those suitable for describing similarity among the species.[*] This is because the species set is usually a fixed entity: if we study vascular plants, then all the vascular plants in any plot are recorded – or at least we believe we have recorded them. But the sample set is usually just a (random) selection, something which is not fixed. Accordingly, the similarity of (the vascular flora of) two samples is a

---

[*] By **similarity of samples**, we mean the similarity in their species composition. By **similarity of species**, we mean the similarity of their distribution among the samples, which presumably reflects the similarity of their ecological behaviour (those are often called species association measures).

meaningful characteristic, regardless of the sample set with which we are working.[+]

On the other hand, similarity of two species has a meaning only within some data sets; for example, *Oxycoccus quadripetalus* and *Carex magellanica* might exhibit a high similarity within a landscape (because they are found together on peat bogs), whereas they will have a low similarity within a peat bog (because *Oxycoccus* is found on hummocks and *Carex* in hollows). The set of analysed samples is considered a random selection of all possible samples in the studied area (in classical statistical terminology, a sample of a population of sampling units). The species similarity in the analysed data set can be considered an estimate of the species similarity in the studied area. Therefore, standard errors of estimate can be (and in some cases have been) derived for the species similarities (see, e.g. Pielou 1977, p. 210). Naturally, when the studied area is a peat bog, the results will be different from those when the studied area is a mountain complex.

Also, there is a reasonable null model for species similarity: the species occur independently of each other (this corresponds to the null hypothesis for a 2×2 contingency table or for a correlation analysis). Consequently, many of the measures of species similarity are centred around zero, with the zero value corresponding to species independence (all the coefficients mentioned in the following text). For similarities of samples, there is no similar null model.

6.1.   Similarity measures for presence–absence data

There are two classical similarity indices that are used in vegetation studies for evaluating the **similarity of samples**: the Jaccard coefficient (Jaccard 1901) and the Sörensen (Sörensen 1948) coefficient. To be able to present their formulae, we first define values of a, b, c and d using a 2×2 frequency table:

		Species in sample B	
		present	absent
Species in	present	a	b
sample A	absent	c	d

Therefore, a is the number of species present in both samples being compared, b is the number of species present in sample A only, c is the number of

---

[+] When the sample set is extended, new species might appear in the whole data set, but all of them are absent from the samples already present in the original set.

species present in sample B only, and d is the number of species absent from both samples. The **Sörensen coefficient** is then defined as

$$S = \frac{2a}{2a + b + c}$$

and the **Jaccard coefficient** as

$$J = \frac{a}{a + b + c}$$

Although there are some differences between these two coefficients, the results based on them are roughly comparable. For both coefficients, the similarity of identical samples equals 1 and the similarity of samples sharing no species is 0. When we need to express the dissimilarity of samples, usually $1 - S$, or $1 - J$ is used.[X] Note that the above coefficients do not use the d value. The number of species absent from both samples does not convey any ecologically interesting information, except in some specific cases.

For calculating the **similarity of species**, a, b, c and d are defined as follows:

		Species B present	Species B absent
		present	absent
Species A	present	a	b
	absent	c	d

where a is the number of samples where both species are present, b is the number of samples only supporting species A, c is the number of samples with species B only, and d is the number of samples where neither of the two species is present. As a measure of species similarity, we usually employ the measures of 'association' in a 2×2 table. The most useful ones are probably the V coefficient (based on the classical chi-square statistic for 2×2 tables):

$$V = \frac{ad - bc}{\sqrt{(a + b)(c + d)(a + c)(b + d)}}$$

and the Q coefficient:

$$Q = \frac{ad - bc}{ad + bc}$$

Values of both V and Q range from -1 to $+1$, with 0 corresponding to the case where the two species are found in common exactly at the frequency corresponding to the model of their independence (i.e. where the probability of

---

[X] But you should use square-roots of these values when you need to obtain dissimilarities (distances) with metric properties.

common occurrence P_{ij} is a product of the independent probabilities of occurrences of the species: $P_{ij} = P_i \cdot P_j$). Positive values mean a positive association, negative values imply a negative association.

The difference between the V and Q measures is in their distinguishing between complete and absolute association (see Pielou 1977). **Complete association** is the maximum (minimum) possible association under given species frequencies. This means that for a complete positive association, it is sufficient that either b or c is zero (i.e. the rarer of the species is always found in the samples where the more common species is present). Similarly, for complete negative association, a or d must be zero. **Absolute** positive association occurs when both species have the same frequency and are always found together (i.e. $b = c = 0$). Absolute negative association means that each of the samples contains just one of the two species (i.e. $a = d = 0$). V is also called the **point correlation coefficient** because its value is equal to the value of the Pearson correlation coefficient of the two species, when we use the value of 1 for the presence of species and 0 for its absence. This is a very useful property, because qualitative data are usually coded this way and the classical Pearson correlation coefficient is readily available in virtually all computer programs.[*]

Note that, unlike the similarity of samples, the d value is absolutely necessary for the similarity of species (contrary to recommendations of, for example, Ludwig and Reynolds 1988, and others). We illustrate this by an example:

Table A

		Species B	
		present	absent
Species	present	50	50
A	absent	50	1000

Table B

		Species B	
		present	absent
Species	present	50	50
A	absent	50	5

In Table A, the two species, both with a low frequency, are found together much more often than expected under the hypothesis of independence (this might be the example of *Carex* and *Oxycoccus* at the landscape scale). Accordingly, both V and Q are positive. On the other side, in Table B, both species have relatively high frequencies, and are found much less in common than expected under the independence hypothesis (e.g. *Carex* and *Oxycoccus* within a

---

[*] However, we cannot trust here the related statistical properties of the Pearson correlation coefficient, such as confidence intervals, statistical significance, etc., because their derivation is based on the assumption of normality. For the statistical properties of V and Q coefficients, see Pielou (1977, p. 210).

peat bog). In this case, both V and Q are negative. Clearly, the species similarity has a meaning only within the data set we are analysing. Therefore, the similarity indices that ignore the d value are completely meaningless. Our experience shows that their value is determined mainly by the frequencies of the two species being compared. For further reading about species association, Pielou (1977) is recommended.

6.2.   Similarity measures for quantitative data

Transformation and standardization

When we use the (dis)similarities based on species presence–absence, the information on species quantity in a sample is lost. From an ecological point of view, it may be very important to know whether the species is a subordinate species found in a single specimen or whether it is a dominant species in the community. There are various quantitative measures of species representation (sometimes generally called importance values): number of individuals, biomass, cover, or frequency in subunits. Often the quantity is estimated on some semi-quantitative scale (e.g. the Braun–Blanquet scale for data about vegetation). The quantitative data are sometimes transformed and/or standardized (see Sections 1.8 and 4.4). **Transformation** is an algebraic function $Y'_{ik} = f(Y_{ik})$ which is applied to each value independently of the other values. **Standardization** is done, on the other hand, with respect to either the values of other species in the same sample (standardization by samples) or the values of the same species in the other samples (standardization by species).

In vegetation ecology, the two extremes are presence–absence and non-transformed species cover (similar results would be obtained using non-transformed numbers of individuals, biomass or basal area values). In between, there are measures such as an ordinal transformation of the Braun–Blanquet scale (i.e. replacement of the original scale values r, $+$, $1 \ldots$ with the numbers $1, 2, 3, \ldots$) which roughly corresponds to a logarithmic transformation of the cover data (or log-transformed cover, biomass, etc.). In many vegetation studies, this is a reasonable and ecologically interpretable compromise. Similarly in animal ecology, the two extremes are presence–absence, and numbers of individuals or biomass. The log transformation is again often a reasonable compromise.

Van der Maarel (1979) showed that there is a continuum of transformations that can be approximated by the relation $Y_{\text{transf}} = Y^z$ (see also McCune and Mefford 1999). When $z = 0$ the data values are reduced to presence and absence. As the value of z increases, the final result is increasingly affected by the dominant species. For similar purpose, the Box–Cox transformation (Sokal & Rohlf 1995) could also be used.

We have shown (Lepš & Hadincová 1992) that, when the analysis is based on cover values, ignoring all species in a relevé that comprise less than 5% of the total cover does not significantly affect the results. The results in this case are completely dependent on the dominant species. Because of the typical distribution of species importance values in most ecological communities (i.e. few dominants and many subordinate species, whose counts and/or biomass are much lower than those of the dominant ones), the same will be true for non-transformed counts of individuals or biomass data. A $\log(Y + 1)$ transformation might be used to increase the influence of subordinate species.

The data can also be standardized (and centred) by samples and/or by species. **Centring** ('centering' in CANOCO user interface) means subtraction of the mean so that the resulting variable (species) or sample has a mean of zero. **Standardization** usually means division of each value by the standard deviation. After standardization by the sample (species) norm (square root of the sum of squares of the values), the vector corresponding to the resulting sample or species values has a unit length.[*] We can also standardize individual samples or species by the total of all the values in a sample (species), thus changing the values to percentages.

We should be extremely careful with standardization by species (either with or without centring). The intention of this procedure is to give all the species the same weight. However, the result is often counter-productive, because a species with a low frequency of occurrence might become very influential. If the species is found in one sample only, then all of its quantity is in this sample alone, which makes this sample very different from the others. On the other hand, the species that are found in many samples do not attain, after standardization, a high share in any of them and their effect is relatively small.

When we calculate similarities on data other than species quantities, often each variable has its own scale. In this case, it is necessary to standardize the variables by calculating the *z-scores*. This procedure corresponds to centring and subsequent standardization by the variable norm. A typical example of this comes from classical taxonomy: each object (individual, population) is described by several characteristics, measured in different units (e.g. number of petals, density of hairs, weight of seeds, etc.). When the similarity is calculated from the rough data, the weight of individual variables changes when we change their units – and the final result is completely meaningless.

---

[*] The standardization by sample (or species) norm (also called **normalization**) is equivalent to standardization by the standard deviation value if the sample (or species) vectors have been centred first. When we mention standardization in further text, we refer to standardization by the norm.

The standardization by samples (either by sample norm or by the total) has a clear ecological meaning. If we use it, we are interested only in proportions of species (both for the standardization by totals and by the sample norm). With standardization, two samples containing three species, in the first sample with 50, 20 and 10 individuals and in the second sample with 5, 2 and 1 individual, will be found to be identical.[X] Standardization by the total (i.e. to percentages) is more intuitive; for example, zoologists frequently use those values and call them 'dominance'. The standardization by sample norm is particularly suitable when applied before calculating the Euclidean distance (see below). The standardization by samples should be routinely used when the abundance total in a sample is affected by the sampling effort or other factors that we are not able to control, but which should not be reflected in the calculated similarities.[+]

Similarity or distance between samples

Probably the most common measure of dissimilarity between samples is **Euclidean distance (ED)**. For a data matrix with m species, with the value of the kth species in the ith sample written as Y_{ik}, the ED is the Euclidean distance between the two points:

$$\text{ED}_{1,2} = \sqrt{\sum_{k=1}^{m} (Y_{1k} - Y_{2k})^2}$$

If we consider the samples to be points in multidimensional space, with each dimension corresponding to a species, then ED is the Euclidean distance between the two points in this multidimensional space. ED is a measure of dissimilarity, its value is 0 for completely identical samples and the upper limit (when the samples have no species in common) is determined by the representation of the species in the samples. Consequently, the ED might be relatively low, even for samples that do not share any species, if the abundances of all present species are very low. This need not be considered a drawback, but it must be taken into account in our interpretations. For example, in research on seedling recruitment, the plots with low recruitment can be considered similar to each other in one of the analyses, regardless of their species composition, and this is then correctly reflected by the values of ED without standardization (see Chapter 14). If we are interested mainly in species composition, then standardization by sample norm is recommended. With standardization

---

[X] The differences in ecological interpretations of analyses with and without standardization by sample norm are discussed in Chapter 14.

[+] For example, the number of grasshoppers caught in a net depends on the sampling effort and also on the weather at the time of sampling. If we are not sure that both factors were constant during all the sampling, it is better to standardize the data.

Table 6-1. *Hypothetical table*

Species	Samples											
	1	2	3	4	1t	2t	3t	4t	1n	2n	3n	4n
1	10		5		1		0.33		1		0.58	
2		10	5			1	0.33			1	0.58	
3			5				0.33				0.58	
4				5				0.33				0.58
5				5				0.33				0.58
6				5				0.33				0.58

Hypothetical table with samples 1 and 2 containing one species each and samples 3 and 4, containing three equally abundant species each (for standardized data the actual quantities are not important). Sample 1 has no species in common with sample 2 and sample 3 has no species in common with sample 4. The samples with t in their labels contain values standardized by the total; those with n are samples standardized by sample norm. For samples standardized by total, $ED_{1,2} = 1.41(= \sqrt{2})$, whereas $ED_{3,4} = 0.82$, whereas for samples standardized by sample norm, $ED_{1,2} = ED_{3,4} = 1.41$.

by sample norm (sometimes called **standardized Euclidean distance** or the **chord distance**, Orloci 1967), the upper limit of the dissimilarity is $\sqrt{2}$ (i.e. the distance between two perpendicular vectors of unit length).

Legendre & Gallagher (2001) discuss various data transformations and their effect on the Euclidean distance (sometimes a special name is coined for the ED after a particular data transformation is used).

The use of ED is not recommended with standardization by sample totals. With this standardization, the length of the sample vector decreases with the species richness and, consequently, the samples become (according to ED) more and more similar to each other. We will illustrate this situation by an example in Table 6-1. The decrease in similarity with species richness of the sample is quite pronounced. For example, for two samples sharing no common species, with each sample having 10 equally abundant species, the ED will be only 0.45.

Among the ordination methods, the linear methods (PCA, RDA) reflect the Euclidean distances between the samples (with the corresponding standardization used in the particular analysis).

Another very popular measure of sample similarity is **percentage** (or **proportional**) **similarity**. Percentage similarity of samples 1 and 2 is defined as

$$PS_{1,2} = \frac{2 \sum\limits_{k=1}^{m} \min(Y_{1k}, Y_{2k})}{\sum\limits_{k=1}^{m} (Y_{1k} + Y_{2k})}$$

Its value is 0 if the two samples have no common species, and it is 1.0 for two identical samples (sometimes it is multiplied by 100 to produce values on the percentage scale). If needed, it can be used as the **percentage dissimilarity** PD = 1–PS (or PD = 100–PS, when multiplied by 100), also called the **Bray–Curtis distance** (Bray & Curtis, 1957). When used with presence–absence data (0 = absence, 1 = presence), it is identical to the Sörensen coefficient. If it is used with data standardized by the sample total, then its value is:

$$\text{PS}_{1,2} = \sum_{k=1}^{m} \min(Y'_{1k}, Y'_{2k})$$

with

$$Y'_{ik} = \frac{Y_{ik}}{\sum\limits_{k=1}^{m} Y_{ik}}$$

These resemblance measures are used under a plethora of names (see, for example, Chapter 7 in Legendre & Legendre 1998, or McCune & Mefford 1999).

Chi-squared distance is rarely used explicitly in ecological studies. However, as linear ordination methods reflect Euclidean distance, unimodal (weighted-averaging) ordination methods (CA, DCA, CCA) reflect chi-square distances. In its basic version, the distance between samples 1 and 2 is defined as:

$$\chi^2_{1,2} = \sqrt{\sum_{k=1}^{m} \frac{S_{++}}{S_{+k}} \left[\frac{Y_{1k}}{S_{1+}} - \frac{Y_{2k}}{S_{2+}} \right]^2}$$

where S_{+k} is the total of the kth species values over all samples:

$$S_{+k} = \sum_{i=1}^{n} Y_{ik}$$

S_{i+} is the total of all the species values in the ith sample:

$$S_{i+} = \sum_{k=1}^{m} Y_{ik}$$

and S_{++} is the grand total:

$$S_{++} = \sum_{i=1}^{n} \sum_{k=1}^{m} Y_{ik}$$

The chi-squared distance is similar to Euclidean distance, but it is weighted by the inverse of the species totals. As a consequence, the species with low frequencies are over-emphasized.[*] The chi-square distance value for two

---

[*] This is why the optional 'downweighting of rare species' is available in unimodal methods.

identical samples is 0 and its upper limit depends on the distribution of the species values. Legendre & Gallagher (2001) have shown that the chi-square distance is actually used when the Euclidean distance is calculated after a particular type of data standardization.

Similarity between species

For species similarity based on quantitative data, the classical **Pearson correlation coefficient** is often a reasonable choice. When it is used for qualitative (presence–absence) data, it is identical to the V coefficient. This shows that it is a good measure of covariation for a wide range of species abundance distributions (nevertheless, the critical values for the Pearson correlation coefficient are valid only for the data with an assumed two-dimensional normal distribution). An alternative choice is provided by one of the rank correlation coefficients, Spearman or Kendall coefficients. Values of all the correlation coefficients range from -1 for a deterministic negative dependence to $+1$ for a deterministic positive dependence, with the value of 0 corresponding to independence. Note that both rank correlation coefficients are independent of the standardization by samples.

6.3.   Similarity of samples versus similarity of communities

The above-described sample similarity measures describe just the similarities between two selected samples. It is often expected that the samples are representative of their communities. In plant ecology, this is usually correct: either we believe in the minimal area concept – Moravec (1973) (and consequently the relevés are taken accordingly and we believe that they are good representatives of the respective communities and, therefore, do not care any further) – or we are interested in the variation in species composition at the spatial scale corresponding to the size of our samples (quadrats).

The situation is very different in many insect community studies. There, the individuals in a sample represent a selection (usually not a completely random one) from some larger set of individuals forming the community. This is especially true in the tropics, where species richness is high and not only a fraction of individuals but also a small fraction of taxa is represented in the sample. Even in large samples, many species are represented by single individuals and consequently the probability that the species will be found again in a sample from a similar community is not high. For example, in samples (greater than 2000 individuals) of insects feeding on tree species in Papua New Guinea, over 40% of species were represented by singletons (Novotný & Basset 2000).

Consequently, the similarity of the two random samples taken from the community might be low and depends on the sample size.

There are methods that attempt to measure the similarity of the sampled communities, and so to adjust for the size of the compared samples. One such measure is the **normalized expected species shared** index (NESS, Grassle & Smith 1976). It is calculated as the expected number of species in common between two random subsamples of a certain size drawn from the two compared larger samples without replacement, normalized (divided) by the average of the expected number of shared species in the two subsamples taken randomly from the first sample and in two subsamples taken randomly from the second sample. These measures, however, are not computationally simple and are not available in common statistical packages. The Morisita index (1959), popular in zoological studies, is a special case of the NESS index. It measures the probability that two individuals selected randomly from *different* subsamples will be the same species. This probability is standardized by the probability that two individuals selected randomly from the *same* subsample will be the same species. It is calculated as:

$$\text{Morisita}_{1,2} = \frac{2 \sum_{k=1}^{m} Y_{1k} Y_{2k}}{(\lambda_1 + \lambda_2) S_{1+} S_{2+}}$$

where

$$\lambda_i = \frac{\sum_{k=1}^{m} Y_{ik}(Y_{ik} - 1)}{S_{i+}(S_{i+} - 1)}$$

The index can be used for numbers of individuals only, so:

$$S_{i+} = \sum_{k=1}^{m} Y_{ik}$$

is the total number of individuals in the ith sample.

6.4.   Principal coordinates analysis

Principal coordinates analysis (PCoA or PCO, also called metric multi-dimensional scaling, and described by Gower 1966) is a multivariate method that attempts to represent in the Cartesian coordinate system a configuration of n objects (samples), defined by a $n \times n$ matrix of dissimilarities (distances) or similarities between these objects. Unlike the principal component analysis (PCA), which can also be defined in similar way, PCoA is able to represent a wide range of similarity (or distance) measures in the space of principal coordinates. Similar to PCA, the first axis (principal coordinate) of the PCoA solution represents the true distances in the matrix in the best possible way as can be achieved with one dimension. As in PCA, each axis has an eigenvalue that indicates its importance. Therefore, starting from the first axis, the importance of principal coordinates decreases in the order of their decreasing eigenvalues.

There is, however, one additional 'problem', which often occurs in the PCoA solution. Not all the (dis)similarity measures can be fully represented in the Euclidean space. If this happens, PCoA produces one or more axes with negative eigenvalues. Because the eigenvalues represent the variance explained on the corresponding principal coordinates, the scores of samples (objects) on those axes are not real, but rather complex numbers (with an imaginary component). These cannot be plotted, due to their non-Euclidean properties. Consequently, the sum of absolute values of such negative eigenvalues related to the sum of absolute values of all the eigenvalues represents the distortion of the original distance matrix properties caused by projection into Euclidean space. Anyway, in the standard applications of PCoA, we do not typically plot all the principal coordinates corresponding to the positive eigenvalues. Rather, only the first few axes are interpreted, similar to other ordination methods.

To eliminate the presence of negative eigenvalues, at least two different correction methods have been developed. These methods adjust the original distance matrix to enable its full embedding in the Euclidean space of PCoA. Instead of using such non-intuitive adjustments, we prefer either to choose a different (dis)similarity measure or to perform its simple, monotonic transformation. For example, when the distances are based on a complement-to-1 (i.e. 1 minus the coefficient value) of a non-metric similarity coefficient (such as the Sörensen coefficient, see Section 6.1), then the square-root transformation of such distance has the required metric properties.

To calculate PCoA in the CANOCO program, you can use the PrCoord program, available from version 4.5. You can use either a pre-computed matrix of distances or you can specify a data file in CANOCO format and select one of the distances that PrCoord is able to calculate. PrCoord produces a new data file (in CANOCO format) which contains sample coordinates on all PCoA axes with positive eigenvalues. This format is handy for using the PCoA results to calculate the constrained PCoA (the distance-based RDA, see Section 6.6) with CANOCO, but you can also use this file to display just the PCoA results with the CanoDraw program. You must use the file produced by PrCoord as the species data file in the CANOCO analysis and specify the principal component analysis (PCA) method, with scaling of scores focused on inter-sample distances, no post-transformation of species scores and centring by species only.

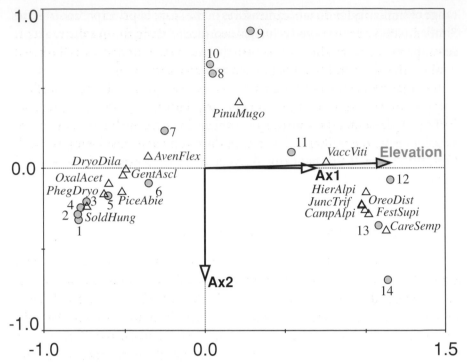

Figure 6-1. First two axes of PCoA with passively projected species centroids (shown as empty triangles), altitude variable (shown as an arrow with shaded head), and position of individual samples (filled circles, labelled with sample numbers). The arrows for the original scores on PCoA axes are also displayed (the first two axes, labelled Ax1 and Ax2).

We will illustrate the application of PCoA using the Tatry data set, described in detail in Section 7.1. To calculate PCoA, start the PrCoord program, specify the *tatry.spe* file as the input datafile, with no transformations, Bray–Curtis distance, and *Do not use* treatment of negative eigenvalues. For example, specify tat_pcoa.dta for the *Output file*. After you click the *Calculate* button, PrCoord runs the analysis and reports that there are 14 principal coordinates (the dimensionality is limited here by the number of samples): 10 first coordinates have positive eigenvalues, one has an eigenvalue equal to zero, and then there are three coordinates with negative eigenvalues. Also note the dominating extent of variation explained on the first axis ($\lambda_1 = 1.226$), compared with the second axis ($\lambda_2 = 0.434$).

Now create a new project in Canoco for Windows, specifying the *tat_pcoa.dta* file as species data. You can use the original species composition data (*tatry.spe*) as supplementary variables, if you want to project them *post hoc* into the PCoA space. The resulting diagram is shown in Figure 6-1.

Here we specified both *tatry.spe* and *tatry.env* as supplementary data, and we assumed a unimodal response of individual species along the gradient, so that the supplementary variables representing plant species are shown as 'nominal variables' (i.e. at centroids of the samples in which they occur).

Note that the original PCoA axes are precisely aligned with the axes of the intermediate PCA and that this data set has probably just one interpretable, altitudinal gradient, represented by the first principal coordinate. This is also documented by the artificial character of the second axis, displaying the arch effect. See Section 7.1 for additional ecological interpretation of the revealed patterns.

6.5.   Non-metric multidimensional scaling

The method of ordination can be formulated in several ways (see Section 3.5). A very intuitive one is finding a configuration in the ordination space in which distances between samples in this space best correspond to dissimilarities of their species composition. Generally, n objects could be ordinated in $n - 1$ dimensional space without any distortion of dissimilarities. The need of clear visual presentations, however, dictates the necessity of reduction of dimensionality (two-dimensional ordination diagrams are most often presented in journals).

This reduction could be done in a metric way, by explicitly looking for projection rules (using principal coordinates analysis, see previous section), or in a non-metric way, by non-metric multidimensional scaling (NMDS, Cox & Cox 1994; Kruskal 1964; Shepard 1962). This method analyses the matrix of dissimilarities between n objects (i.e. samples) and aims to find a configuration of these objects in k-dimensional ordination space (k is *a priori* determined), so that those distances in ordination space correspond to dissimilarities. A statistic termed 'stress' is designed to measure the 'lack of fit' between distances in ordination space and dissimilarities:

$$\text{stress} = \Sigma[d_{ij} - f(\delta_{ij})]^2$$

where d_{ij} is the distance between sample points in the ordination diagram, δ_{ij} is the dissimilarity in the original matrix of distances, calculated from the data, and $f()$ is a non-metric monotonous transformation. In this way, the 'correspondence' is defined in a non-metric way, so that the method reproduces the general rank-ordering of dissimilarities (not exactly the dissimilarity values). The algorithm re-shuffles the objects in the ordination space to minimize the stress. The algorithm is necessarily an iterative one, and its convergence depends on the initial configuration; also, the global minimum is not always achieved and, consequently, it is worthwhile trying various initial configurations. The NMDS method is now available in most general statistical packages, as well as in specialized multivariate packages. The implementation details differ among these programs (for example, it is possible to start from various initial configurations and automatically select the result with lowest stress).

In contrast to the other ordination method, the input data (in both metric and non-metric multidimensional scaling) are not the original samples by species table, but a matrix of similarities between objects. Consequently, virtually any measure of (dis)similarity could be used.[*] Unlike other methods, in NMDS the number of ordination axes (i.e. the dimensionality of the ordination space) is given *a priori*. In real applications, various dimensionalities (number of axes, k) are tried, and the 'proper' number of dimensions is decided according to the 'quality' of resulting models. The quality is measured by the stress of the resulting configuration. As the method is normally used as an indirect ordination, the interpretability of the results is usually one of the criteria. As with the other indirect methods, the axes could be *a posteriori* interpreted using measured environmental variables, which could be *a posteriori* projected to the ordination diagram.

We will demonstrate the use of this method with the Statistica program, on the example data from the elevation gradient in the Tatry Mts., described in Section 7.1. During the classification, you save the dissimilarity matrix in file *tatrydis.sta* (see Section 7.3), and use this matrix now as the input data (or you can calculate any (dis)similarity matrix, e.g. in Excel, and enter it according to the rules for (dis)similarity matrices in Statistica). The startup panel is very simple: you have to specify that you want to analyse all the variables (which implies all the samples, in our case) and the number of dimensions to be used (we will use two in this example). The other options are related to the details of the numerical procedure, and the default values probably need not be changed. The starting configuration will be *Standard G-L*, which means that the starting configuration is calculated by using PCA. Then you could follow the numerical procedure (particularly important when there are problems with convergence). The most important result for us is, however, the final configuration in the ordination space. Even when you use more than two dimensions, graph the final configuration as two-dimensional graphs, combining the axes as needed. The final result is shown in Figure 6-2.

---

[*] Recall that this matrix is implicitly determined by selection of linear/unimodal model in the response-model-based ordinations.

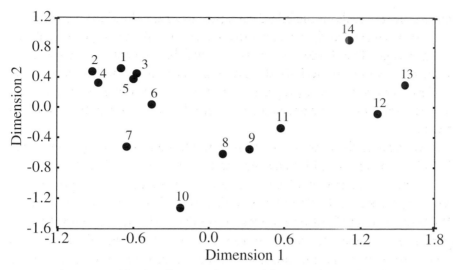

Figure 6-2. Ordination diagram of NMDS of the Tatry data set, based on Euclidean distances.

You can see that the method reasonably reconstructed the elevation gradient: the first axis is strongly correlated with the elevation. The second axis is difficult to interpret, and is clearly a quadratic function of the first axis – even the NMDS do not escape the arch effect.

NMDS is, however, used more often with dissimilarity measures other than Euclidean distance. We applied it to the matrix of percent dissimilarities. The matrix was calculated using a macro in Excel, file *persim.xls* – see Chapter 7. The results of NMDS based on percent dissimilarities (Figure 6-3) are (in this case) very similar to those based on the Euclidean distances – including the quadratic dependence of the second axis on the first.

The specialized programs provide more advanced options for NMDS than the Statistica program. For example, the PCORD program provides a procedure assisting you with the selection of an optimum dimensionality for the NMDS solution.

6.6.   Constrained principal coordinates analysis (db-RDA)

A constrained PCoA method, representing a canonical form for the analysis of a matrix of (dis)similarities, was proposed by Legendre & Anderson (1999) under the name **distance-based RDA** (db-RDA). To perform this type of analysis, the analysis of principal coordinates must be performed on the matrix of sample (dis)similarities and the resulting sample coordinates (using all the axes with positive eigenvalues, after optional correction for negative

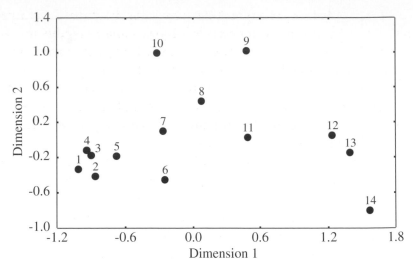

Figure 6-3. Ordination diagram of NMDS of the Tatry data set, based on percent dissimilarities.

eigenvalues, see Section 6.4) are entered as species data into a redundancy analysis (RDA).

This method enables us to test hypotheses about the effects of explanatory variables on community composition, while (almost) freely selecting the measure of (dis)similarity among the collected samples. Distance-based RDA can use covariables and also reflect non-random sampling or the experimental design during Monte Carlo permutation tests.

Note, however, that we pay for this flexibility by losing the model of species relation with the (constrained) ordination axes, which is fruitfully used when interpreting the standard constrained ordinations (RDA or CCA). While we can project selected species *a posteriori* into the db-RDA diagram, there is no implied model of species response. Therefore, our decision to show species as arrows (implying linear change of species abundances across the db-RDA space) or as centroids (implying unimodal models) is an arbitrary choice, with no guaranteed coherence with the selected (dis)similarity coefficient.

6.7.   Mantel test

In some studies, we are interested in the relationship of two similarity or distance matrices. For example, we have n individuals in a plant population, each individual being characterized by its genetic constitution (e.g. allelic composition, determined by methods of molecular biology such as PCR

(polymerase chain reaction) or RFLP (restriction fragment length polymor-
phism)), and also by its position at a locality (as X and Y coordinates). We want
to test whether the distance between plants reflects genetic similarity. We can
use the X and Y variables as predictors and the allelic composition as a response
matrix and apply some constrained ordination. However, this would test for a
spatial trend in the genetic composition, and would show no relationship if the
population were composed of spatially restricted groups of genetically similar
individuals scattered over the locality.

An alternative approach is to calculate two matrices: the matrix of genetic
similarities and the matrix of physical distances between the individuals. We
can directly calculate the classical (Pearson) correlation coefficient (r) between
the corresponding values of physical distance and genetic similarity (or we can
use a regression of genetic similarity on the physical distance). Such analyses
provide reasonable measures of how closely the two matrices are related (r) or
how much of the genetic similarity is explained by the physical proximity (coefficient of determination R^2). However, the parametric test of significance for
r (or the estimate of standard error of r) will not be correct, because both are
based on the assumption that the observations are independent, which is not
true: the number of degrees of freedom is greatly inflated, as we have $n(n-1)/2$
pairs (distances) that enter the analysis, but only n independent observations.
If there is an individual which is genetically different from the others, all its
$n-1$ similarities to other individuals will be very low, all based on a single
independent observation.

The best solution is to estimate the distribution of the test statistics (e.g. the
Pearson linear correlation r) under the null hypothesis using Monte Carlo permutations. If there is no relationship between the spatial distance and genetic
similarity, then the locations of individuals can be randomly permuted, and
all the spatial arrangements are equally likely. Consequently, in the permutation, not the individual similarities (distances), but the identity of individuals
is permuted in one of the matrices (it does not matter which one). In this way,
the internal dependencies within each of the matrices are conserved in all the
permutations, and only the relationship between the spatial coordinates and
genetic properties is randomized.

One-tailed or two-tailed tests could be used, depending on the nature of
the null hypothesis. In the above example, we would very probably use the
one-tailed test, i.e. the null hypothesis would be that the genetic similarity is either independent of, or increases with, physical distance, and the
alternative hypothesis would be that the genetic similarity decreases with
the distance (we can hardly imagine mechanisms that cause the genetic similarity to increase with physical distance between individuals). Because the

alternative hypothesis suggests negative r values, the estimate of type I error probability is:

$$P = \frac{n_x + 1}{N + 1}$$

where n_x is number of simulations where $r < r_{\text{data}}$; for the two-tailed test, n_x is the number of simulations where $|r| > |r_{\text{data}}|$.

In practical calculations, the sum of products (which is quicker to calculate) is used instead of r (which is, on the other hand, more intuitive). However, it can be shown that the results are identical. Also, instead of permutations, an asymptotic approximation can be used.

One of the two compared distance (or similarity) matrices can also be constructed using arbitrary values, reflecting the alternative hypothesis in the test. For example, when studying differences between two or more groups of objects or samples, we can use a distance matrix, where the within-group distances are set to 0 and the between-group distances are set to 1. See Section 10.5.1, of Legendre & Legendre (1998), for further discussion.

The other examples of the use of the Mantel test in ecology are as follows.

1. It can be expected that species with similar phenology compete more than species with different phenology. Species that compete more can be expected to have negative 'inter-specific association', i.e. they are found together less often than expected, when recorded on a very small scale in a homogeneous habitat (because of competitive exclusion). Two matrices were calculated, the matrix of phenological similarity of species and the matrix of species associations, and their relationship was evaluated by the Mantel test (Lepš & Buriánek 1990). No relation was found in this study.

2. In a tropical forest, a herbivore community of various woody plant species was characterized by all the insect individuals (determined to the species) collected on various individuals of the woody species in the course of one year. Then, the abundance of all the investigated woody species in about 100 quadrats was recorded. The two matrices – matrix of similarities of woody species' herbivore communities, and matrix of 'inter-specific associations' of woody species, based on the quadrat study – were compared using the Mantel test. It was shown that species with similar distribution also had similar herbivore communities, even when the relationship was weak (Lepš et al. 2001).

The Mantel test is now available in specialized statistical packages, such as the PCORD or PRIMER programs (see Appendix C). Note that the Mantel test is also sometimes applied to problems where the two matrices are calculated from the species composition data table and the table with explanatory variables. Therefore, the various constrained ordination methods (CCA, RDA, db-RDA) can also be applied to the same problem, with the extra advantage of visualizing the relation between the ecological community and the environment in terms of individual species.

Classification methods

The aim of classification is to obtain groups of objects (samples, species) that are internally homogeneous and distinct from the other groups. When the species are classified, the **homogeneity** can be interpreted as their similar ecological behaviour, as reflected by the similarity of their distributions. The classification methods are usually categorized as in Figure 7-1.

Historically, numerical classifications were considered an objective alternative to subjective classifications (such as the Zürich-Montpellier phytosociological system, Mueller-Dombois & Ellenberg 1974). They are indeed 'objective' by their reproducibility, i.e. getting identical results with identical inputs (in classical phytosociology, the classification methods obviously provide varying results). But also in numerical classifications, the methodological choices are highly subjective and affect the final result.

7.1.   Sample data set

The various possibilities of data classification will be demonstrated using vegetation data of 14 relevés from an altitudinal transect in Nízké Tatry Mts, Slovakia. Relevé 1 was recorded at an altitude of 1200 metres above sea level (m a.s.l.), relevé 14 at 1830 m a.s.l. Relevés were recorded using the Braun–Blanquet scale (r, +, 1–5, see Mueller-Dombois & Ellenberg 1974). For calculations, the scale was converted into numbers 1 to 7 (ordinal transformation, Van der Maarel 1979). Data were entered as a classical vegetation data table (file *tatry.xls*) and further imported (using the WCanoImp program) into the condensed Cornell (CANOCO) format (file *tatry.spe*), to enable use of the CANOCO and TWINSPAN programs. The data were also imported into a Statistica file (file *tatry.sta*).

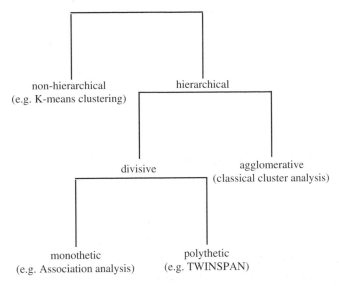

Figure 7-1. Types of classification methods.

First, we look at the similarity structure displayed by the DCA ordination method (see Section 3.7). The resulting diagram with species and samples is shown in Figure 7-2.

The graph (Figure 7-2) demonstrates that there is linear variation in the data, corresponding to the altitudinal gradient. Samples 1 to 5 are from the spruce forest (characterized by plant species *Picea abies*, *Dryopteris dilatata*, *Avenella flexuosa*, etc.), and samples 12 to 14 are from typical alpine grassland (e.g. *Oreochloa disticha*), and there is also a dwarf-pine (*Pinus mugo*) zone in between (with *Calamagrostis villosa*, *Vaccinium myrtillus*, see Figure 7-3). Now we will explore how the particular classification methods divide this gradient into vegetation types and also demonstrate how to perform the individual analyses.

7.2.   Non-hierarchical classification (K-means clustering)

The goal of this method is to form a pre-determined number of groups (clusters). The groups should be internally homogeneous and different from each other. All the groups are on the same level, there is no hierarchy. Here, we will use K-means clustering as a representative of non-hierarchical classifications.

For the computation, an iterative relocation procedure is applied. The procedure starts with k (desired number of) clusters, and then moves the objects to minimize the within-cluster variability and maximize the between-cluster

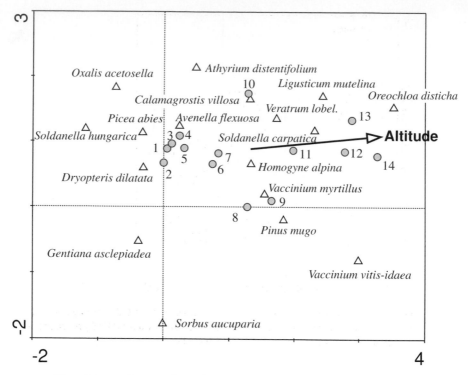

Figure 7-2. Species-samples ordination diagram with results of DCA of the altitudinal transect in Tatry Mts (sample altitude is passively projected into the diagram). Only the species with the highest weight (the most frequent ones) were selected for display.

variability. When the clusters are different, then ANOVA for (at least some) species shows significant results, so the procedure can be thought of as 'ANOVA in reverse', i.e. forming groups of samples to achieve the most significant differences in ANOVA for the maximum number of species (variables).

The relocation procedure stops when no move of an object (sample) improves the criterion. You should be aware that you might get a local extreme with this algorithm, and you can never be sure that this is the global extreme. It is advisable to start with different initial clusters and check if the results are the same for all of them.

We will use the Statistica program (procedure *K-means clustering*). In Statistica, choose the *Cluster Analysis* procedure and select *K-means clustering*. The variables to be used in clustering must be selected. To do so, click on the *Variables* button and select the variables you want to use. Alternatively, you can click on the *Variables* button and select only a subset and calculate the classification using selected species only (just the herbs, for example).

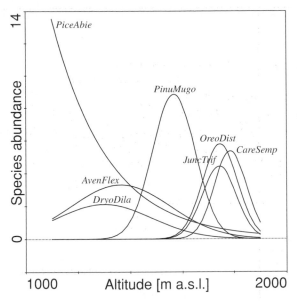

Figure 7-3. Response curves of the important species on the elevation gradient, fitted in CanoDraw by a second-order polynomial predictor (GLM procedure) – see Chapter 8. m a.s.l. means metres above sea level.

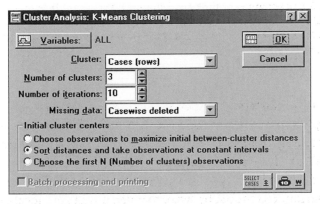

Figure 7-4. Dialog box for specifying K-means clustering in the Statistica for Windows program.

In the next panel, select for *Members of each cluster & distances.* Here we learn that samples 1–6 are in the first cluster, samples 7–11 are in the second cluster, and the samples 12–14 are in the third one. The fact that the samples in each cluster are contiguous in numbering is caused by the linear character of the variation in our data. For other data sets, we can get one group containing, for example, samples 1, 3 and 5, and the other group containing samples 2 and 4. For each cluster, we see the distances to the centre of the corresponding cluster, e.g. for cluster 1:

Members of Cluster Number 1 (tatry.sta)					
and Distances from Respective Cluster Center					
Cluster contains 6 cases					
	Case No.	Case No.	Case No.	Case No.	Case No.	Case No.
	C_1	C_2	C_3	C_4	C_5	C_6
Distance	0.49637	0.577225	0.456277	0.600805	0.520278	1.017142

The samples seem to form a homogeneous cluster, with the exception of sample 6, being an outlier (similarly, sample 10 is an outlier in cluster 2). The results are in agreement (including the determination of outliers) with the DCA results.

Now, we can select for *Cluster means and Euclidean distances.* The table of the Euclidean distances provides information on the similarity among the individual clusters, and the table of cluster means displays the mean values of the species in the particular clusters.

We can see that the most similar are clusters 1 and 2, the most dissimilar are 1 and 3 (as expected):

Euclidean Distances between Clusters (tatry.sta)			
Distances below diagonal			
Squared distances above diagonal			
	No. 1	No. 2	No. 3
No. 1	0	1.616661	3.178675
No. 2	1.27148	0	2.212292
No. 3	1.782884	1.487377	0

Further, the representation of species in the individual clusters is clear from the table of means:

Cluster Means (tatry.sta)			
	Cluster No. 1	Cluster No. 2	Cluster No. 3
PICEABIE	5.416667	1.6	0
SORBAUCU	1.75	1.4	0
PINUMUGO	0.333333	6.4	1.333333
SALISILE	0	0.8	0

Clearly, *Picea abies* is common in cluster 1 (spruce forests) and missing in cluster 3 (alpine meadows). The dwarf pine, *Pinus mugo*, is rare outside the middle 'krumbholz' zone. Further useful information is provided by the *Analysis of variance*:

	Between SS	df	Within SS	df	F	signif. p
PICEABIE	71.68095	2	21.40831	11	18.41552	0.000309
SORBAUCU	6.3	2	23.575	11	1.469778	0.271826
PINUMUGO	107.6571	2	15.2	11	38.9549	1.02E-05
SALISILE	2.057143	2	12.8	11	0.883929	0.440566
JUNICOMM	6.547619	2	4.666667	11	7.716836	0.00805
VACCMYRT	0.32381	2	20.03333	11	0.0889	0.915588
OXALACET	16.16667	2	19.33333	11	4.599139	0.035353
HOMOALPI	1.414286	2	4.3	11	1.808971	0.209308
SOLDHUNG	18.66666	2	7.333336	11	13.99999	0.000948
AVENFLEX	16.89524	2	4.033331	11	23.03897	0.000117
CALAVILL	0.928572	2	40.875	11	0.124945	0.88378
GENTASCL	7.295238	2	5.633332	11	7.122571	0.010368
DRYODILA	8.914286	2	4.8	11	10.21428	0.003107

For each species, an ANOVA comparing the means in the three clusters is calculated. Note that you should not interpret the p-values as ordinary type I error probabilities, because the clusters were selected to maximize the differences between them. Nevertheless, the p-values provide a useful indication as to which species differ considerably between clusters (or on which species the classification is based). *Picea abies*, *Pinus mugo*, or *Soldanella hungarica* averages differ considerably between the clusters, whereas *Vaccinium myrtillus* averages do not (this species is relatively common along the whole transect).

Program Statistica does not allow standardization, but it might be a reasonable decision to standardize the data by sample norm. If you decide to, you have to carry out the standardization in the spreadsheet, before importing the data into the Statistica program.

7.3.   Hierarchical classifications

In hierarchical classifications, groups are formed that contain subgroups, so there is a hierarchy of levels. When the groups are formed from the bottom (i.e. the method starts with joining the two most similar objects), then the classifications are called **agglomerative**. When the classification starts with division of the whole data set into two groups, which are further split, the classification is called **divisive**. The term **cluster analysis** is often used for agglomerative methods only.

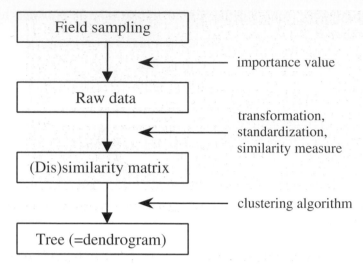

Figure 7-5. The methodological decisions affecting the results of an agglomerative hierarchical classification.

Agglomerative hierarchical classification (cluster analysis)

The aim of these methods is to form a hierarchical classification (i.e. groups containing subgroups) which is usually displayed by a **dendrogram**. The groups are formed 'from the bottom', i.e. the most similar objects are first joined to form the first cluster, which is then considered as a new object, and the joining continues until all the objects are joined in the final cluster, containing all the objects. The procedure has two basic steps: in the first step, the similarity matrix is calculated for all the pairs of objects.* In the second step, the objects are clustered/amalgamated so that, after each amalgamation, the newly formed group is considered to be an object and the similarities of the remaining objects to the newly formed one are recalculated. The individual methods (algorithms) differ in the way they recalculate the similarities.

You should be aware that there are several methodological decisions affecting the final result of a classification (see Figure 7-5).

Agglomerative hierarchical classifications are readily available in most statistical packages. We will demonstrate their use in the Statistica package; however, the rules are similar in most packages. Statistica, similar to other programs, allows for a direct input of the similarity matrix. This is useful because Statistica contains only a limited number of dis/similarity measures. It

---

* The matrix is symmetrical, and on the diagonal there are either zeroes – for dissimilarity – or the maximum possible similarity values.

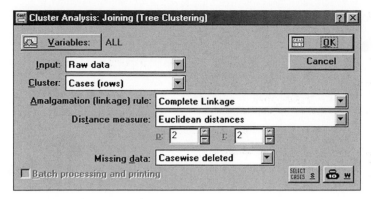

Figure 7-6. Dialog box for hierarchical, agglomerative clustering in Statistica for Windows.

is possible to prepare a simple macro, e.g. in Excel, that will calculate the similarities and then import the similarity matrix into the Statistica program.*

In its basic form, the procedure used in the Statistica program is quite simple:

- Select *Joining (tree clustering)* in the startup panel: for clustering of samples, use the options in Figure 7-6.
- *ALL* the variables are selected (you can also do the analysis with a subset of variables).
- *Raw data* means that the raw data are in your data file, not the similarity matrix.
- We selected the *Complete Linkage* amalgamation (clustering) procedure. There are other possibilities and the decision affects the resulting dendrogram. There are 'shorthand' methods (e.g. *Single Linkage*) in which the distance between the clusters is defined as the distance between the closest points in the two compared clusters. These methods produce dendrograms characterized by a high level of *chaining*. The 'longhand' methods (e.g. *Complete Linkage*) in which distance between two clusters is defined as the distance between the furthest points, tend to produce compact clusters of roughly equal size (Figure 7-7).

---

* You need to prepare the file according to the rules for similarity matrices in Statistica: a symmetrical square matrix, with column names being the same as the row names, and an additional four rows as follows. The first and second rows, which must be called *Means* and *Std. dev.*, respectively, contain averages and standard deviations of variables (necessary only for the correlation and covariance type). The first column of the third row, which is called *No Cases*, contains the number of items on which the similarities are based (this does not affect the results of clustering). And , finally, the first column of the fourth row, which must be called *Matrix*, contains the matrix type: 1 = correlation, 2 = similarities, 3 = dissimilarities, and 4 = covariance.

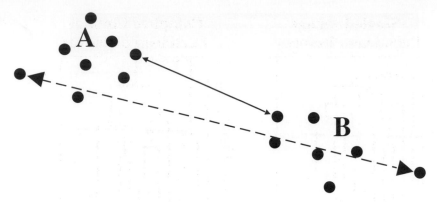

Figure 7-7. Distance between two clusters, A and B, as defined in the single linkage (solid line) and in the complete linkage (dashed line) algorithms.

- After completing the analysis, it is useful to save the distance matrix, using the '*save distance matrix*' button, in a file with the name *tatrydis.sta*. This file can be used for calculating the non-metric multidimensional scaling (NMDS, see Section 6.5).

There are many other methods whose underlying approach usually falls somewhere between the two above-described methods. Among them, the **average linkage** method used to be very popular. However, this term was used inconsistently. The term **unweighted-pair groups method** (UPGMA) describes its most common variant. In ecology, the 'short hand' methods are usually of little use. Compare the resulting dendrograms on the left and right sides of Figure 7-8. The results of the complete linkage algorithm are more ecologically interpretable: when these groups are compared to our external knowledge of the nature of the data, they can be judged as better reflecting the elevation gradient. See Sneath (1966) for a comparison of various clustering methods.

We first selected the *Euclidean distance* as a measure of sample similarity. If you prefer to use the standardized Euclidean distance (chord distance) instead, the data must be standardized before being imported into Statistica. Unlike other software (PCORD software, for example, provides a much wider selection of (dis)similarity measures, McCune & Mefford 1999), the Statistica program has quite a limited selection of options, but you can use its ability to import matrix of similarities from external sources (see above for additional details). We have prepared a simple Visual Basic macro in Excel (file *persim.xls*), which calculates percentage dissimilarities and saves them in a spreadsheet in a form that can be directly imported into the Statistica program (including the names of cases and variables). Figure 7-8 contains, for comparison, classifications based both on Euclidean distances and percentage dissimilarity. The

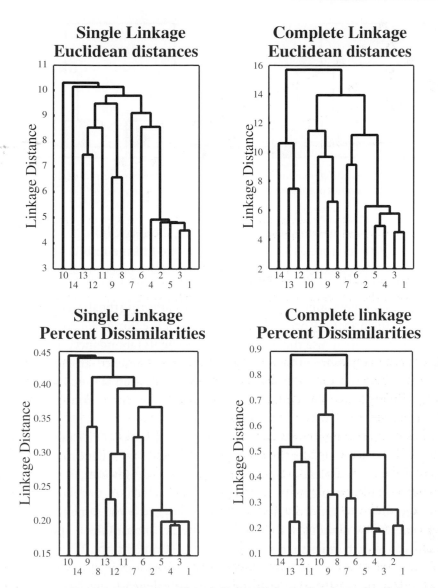

Figure 7-8. Comparison of single linkage (left panel) and complete linkage (right panel) clustering of the samples, based on Euclidean distances (top) and percentage dissimilarities (bottom). Note the higher degree of chaining in the two dendrograms from single linkage (samples 14 and 10 do not belong to any cluster, but are chained to the large cluster containing all the other samples). The complete linkage results can be interpreted more easily.

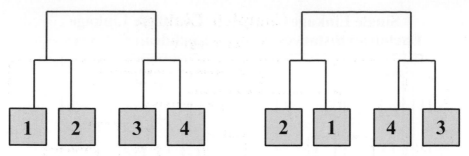

Figure 7-9. Two different dendrograms that represent the same results of cluster analysis. The order of subgroups within a group is arbitrary.

percentage-dissimilarity-based dendrogram using the single linkage method is also difficult to interpret. However, the complete linkage results seem to reflect the elevation gradient slightly better than those based on Euclidean distances.

Also, one of the most important decisions greatly affecting the similarity values is data transformation: according to our experience (Kovář & Lepš 1986), the decision whether to use the original measurements (abundance, biomass or cover), the log-transformation of these values, or the presence–absence data influences the resulting classification more than the selection of the clustering rule. The standardization by samples has a major effect when the sample totals are very different.

For most ordinary agglomerative classifications the same result can be presented by many different dendrograms. What matters in a dendrogram is which entities (e.g. samples or sample groups) are joined. However, it does not matter which of the two groups is on the left and which on the right side. Figure 7-9 shows two dendrograms that represent exactly the same result of cluster analysis. We cannot say, for example, whether the samples 1 and 4 are more similar to each other than samples 2 and 3. Unlike in the TWINSPAN method (described below), the orientation of the subgroups in the agglomerative classification dendrogram is arbitrary (usually depends on the order in which the data are entered) and, therefore, should not be interpreted as being a result of the analysis.

It is interesting to compare the results of the classifications with DCA ordination. The results of detrended correspondence analysis (Figure 7-2) suggest that there is a fairly homogeneous group of samples 1 to 5 (the samples from spruce forest). All the classification methods also recognized this group as being distinct from the remaining samples. DCA suggests that there is rather continuous variation with increasing elevation. The individual classifications differ in the way this continuous variation is split into groups. For example, sample 11 is in an intermediate position between the sub-alpine shrubs and

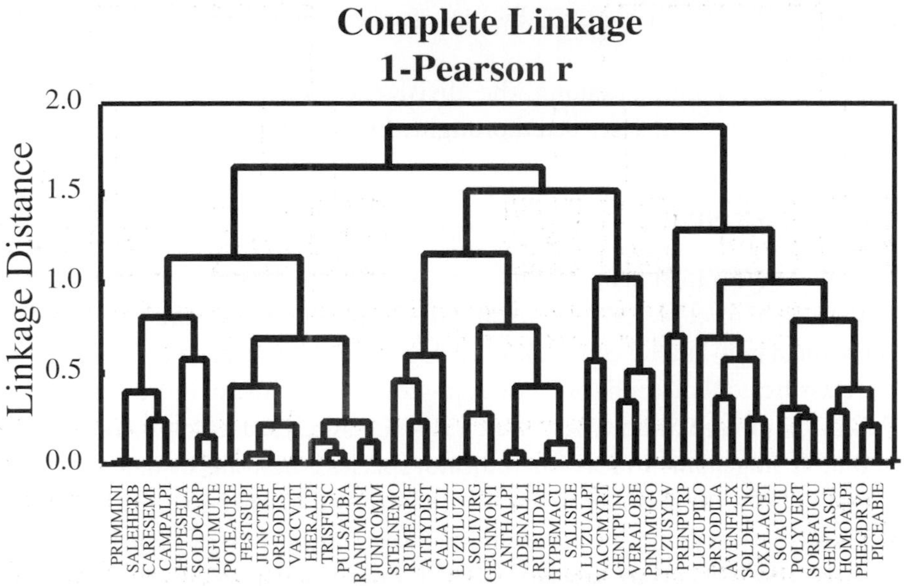

Complete Linkage
1-Pearson r

Figure 7-10. The classification of species. Note that the procedure distinguished a group of alpine grassland species reasonably well (*Primula minima*, *Salix herbacea*, *Carex sempervirens*, *Campanula alpina*) on the left side of the diagram. Similarly, species of the spruce forest are on the right side of the diagram.

alpine meadows in the ordination diagram and even the complete linkage classifications differ in their 'decision': the sample is allocated to the sub-alpine group when the classification is based on Euclidean distances, and to the alpine group when classification is based on percentage similarity.

You can also perform cluster analysis of variables (i.e. of the species in our example). In this case, the correlation coefficient will probably be a reasonable measure of species distributional similarity (to convert it to dissimilarity, $1-r$ is used). Note that the appropriate similarity measures differ according to whether we cluster samples or species. The results of the clustering of species are displayed in Figure 7-10.

Divisive classifications

In **divisive classifications,** the whole set of objects is divided 'from the top': first, the whole data set is divided into two parts, each part is then considered separately and is divided further. When the division is based on a single attribute (i.e. on a single species), the classification is called **monothetic**, when based on multiple species, the classification is **polythetic**. The significance of monothetic methods is mostly a historical one. The classical 'association analysis' was a monothetic method (Williams & Lambert 1959). The advantage of the divisive methods is that each division is accompanied by a rule used for the division, e.g. by a set of species typical for either part of the dichotomy.

The most popular among the divisive classification methods is the TWINSPAN method, described thoroughly in the following section.

7.4.   TWINSPAN

The TWINSPAN method (from Two Way INdicator SPecies ANalysis, Hill 1979; Hill et al. 1975) is a very popular method (and the program has the same name) among community ecologists and it was partially inspired by the classificatory methods of classical phytosociology (use of indicators for the definition of vegetation types). We treat it separately, based on its popularity among ecologists. The idea of an indicator species is basically a qualitative one. Consequently, the method works with qualitative data only. In order not to lose the information about the species abundances, the concepts of **pseudo-species** and pseudo-species **cut levels** were introduced. Each species can be represented by several pseudo-species, depending on its quantity in the sample. A pseudo-species is present if the species quantity exceeds the corresponding cut level. Imagine that we selected the following pseudo-species cut levels: 0, 1, 5 and 20. Then the original table is translated to the table used by TWINSPAN as follows:

	Species	Sample 1	Sample 2
Original table	*Cirsium oleraceum*	0	1
	Glechoma hederacea	6	0
	Juncus tenuis	15	25
Table with pseudo-species used in TWINSPAN	Cirsoler1	0	1
	Glechede1	1	0
	Glechede2	1	0
	Junctenu1	1	1
	Junctenu2	1	1
	Junctenu3	1	1
	Junctenu4	0	1

In this way, quantitative data are translated into qualitative (presence–absence) data.

In TWINSPAN, the dichotomy (division) is constructed on the basis of a *correspondence analysis* (CA) ordination. The ordination is calculated and the

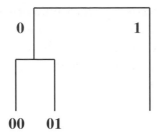

Figure 7-11. Sample splits in a TWINSPAN classification.

samples are divided into the left (negative) side and the right (positive) side of the dichotomy according to their score on the first CA axis. The axis is divided at the centre of gravity (centroid). However, usually there are many samples near the centre of gravity. Consequently, many samples are close to the border, and their classification would depend on many factors. Therefore, a new ordination is constructed, which gives a higher weight to the 'preferentials', i.e. the species preferring one or the other side of the dichotomy. The algorithm is rather complicated, but the aim is to get a polarized ordination, i.e. an ordination where most of the samples are not positioned close to the centre of gravity. Therefore, the classification of the samples is not based so much on the species common to both parts of the dichotomy, but mostly on the species typical of one part of the dichotomy and consequently (in concordance with phytosociological tradition) these species can be considered to be good indicators of particular ecological conditions.

In the first division, the polarity (i.e. which part of the dichotomy will be negative and which positive) is arbitrarily set. In the subsequent divisions, polarity is determined according to the similarity of dichotomy parts to the 'sister' group in a higher-level division. For example, in the dendrogram in Figure 7-11, group 01 is more similar to group 1 than group 00 is to group 1. A result of this process is that the samples are ordered in the final table, and we get a table that is similar to an ordered phytosociological table.

Also, the determined 'rule' (i.e. the set of species, that were important for a particular division) is printed by the program at each step. This greatly increases the interpretability of the final results. The classification of samples is complemented by a classification of species and the final table is based on this two-way classification.

TWINSPAN analysis of the Tatry samples

In the following text, we will demonstrate the use of TWINSPAN for the analysis of 14 relevés from the altitudinal transect that we used in previous sections. TWINSPAN is useful for large data sets, but the small data set is used here for simplicity. We used the file *tatry.spe*, i.e. the file imported using the WCanoImp program into the Cornell-condensed (CANOCO) format. In our sample analysis, we asked for the long output to show what information can be obtained from the program (usually, only short output is requested – even the short output is pretty long).

First, the headings are printed and the program lists the options selected. The important ones are:

```
Cut levels:
        .00   1.00   2.00   3.00   4.00   5.00
```

Because the data were converted into numeric values using ordinal transformation, there is no reason to further 'downweight' the higher values. When the data are in the form of estimated cover, it is reasonable to use the default cut levels, i.e. 0 2 5 ....The cut levels 0 1 10 100 1000 give results corresponding to a logarithmic transformation and are useful when the data are numbers of individuals, differing in order of magnitude.

From further options note the following (all have their default values):

```
1. Minimum group size for division:                5
2. Maximum number of indicators per division:      7
3. Maximum number of species in final tabulation: 100
4. Maximum level of divisions:                     6
```

Option **1** means that groups containing fewer than 5 relevés are terminal, i.e. they are not further divided. For small data sets, it might be reasonable to decrease this value to 4.

Option **2** represents the number of indicators per division; usually, the default value is a reasonable choice.

Option **3**: if you have more than 100 species, only the 100 most common species will be included in the final tabulation. You can increase the value, if necessary.

Option **4** represents an alternative way to control the divisions (controlling by the size of the group using option **1** is a better solution). For a data set of reasonable size, the value 6 is usually sufficient.

The program output starts with the description of first division:

```
DIVISION    1   (N= 14)            I.E. GROUP *
  Eigenvalue  .565 at iteration    1
  INDICATORS, together with their SIGN
  Oreo Dist1(+)
```

The indicator for the first division is *Oreochloa disticha* (the 1 at the end means that the first pseudospecies cut level was used, i.e. the presence of the species is enough for the indicator to be considered present).

```
Maximum indicator score for negative group  0    Minimum indicator score for
positive group    1

 Items in NEGATIVE group    2   (N=  10)        i.e. group *0
Samp0001  Samp0002  Samp0003  Samp0004  Samp0005  Samp0006    Samp0007  Samp0008
Samp0009 Samp0010

 BORDERLINE negatives     (N=    1)
 Samp0009

 Items in POSITIVE group   3   (N=    4)        i.e. group *1
 Samp0011  Samp0012  Samp0013  Samp0014
```

The output displayed above represents the division of samples. Note the warning for sample 9 that this sample was on the border between the two groups (this warning appears only when you ask for a long output).

Now, the species preferring one side of the dichotomy (preferentials) are listed, with the number of occurrences in each of the groups (e.g. *Picea abies* was present in seven samples from the negative group and in one sample from the positive group). Note that the preferentials are determined with respect to the number of samples in each group and also for each pseudospecies cut level separately. Preferentials are provided in the long output only.

```
NEGATIVE PREFERENTIALS
  Pice Abie1(  7, 1) Sorb Aucu1(  7, 0) Oxal Acet1(  7, 0) Sold Hung1(  5, 0)
  Aven Flex1( 10, 1) Gent Ascl1(  8, 0) Dryo Dila1(  8, 0) Pheg Dryo1(  6, 0)
  Pren Purp1(  2, 0) Poly Vert1(  3, 0) SoAu cJu 1(  3, 0) Luzu Pilo1(  2, 0)
etc...

POSITIVE PREFERENTIALS
  Juni Comm1(  0, 2) Ligu Mute1(  4, 4) Sold Carp1(  2, 4) Ranu Mont1(  1, 2)
  Hupe Sela1(  3, 3) Geun Mont1(  2, 2) Vacc Viti1(  2, 4) Puls Alba1(  0, 2)
  Gent Punc1(  2, 3) Soli Virg1(  1, 1) Luzu Luzu1(  1, 1) Oreo Dist1(  0, 4)
etc...

NON-PREFERENTIALS
  Pinu Mugo1(  5, 2) Vacc Myrt1( 10, 4) Homo Alpi1( 10, 4) Cala Vill1(  8, 3)
  Rume Arif1(  4, 1) Vera Lobe1(  5, 3) Pinu Mugo2(  5, 2) Vacc Myrt2( 10, 4)
  Homo Alpi2( 10, 4) Cala Vill2(  8, 3) Rume Arif2(  4, 1) Vera Lobe2(  5, 3)
etc...
  End of level    1
```

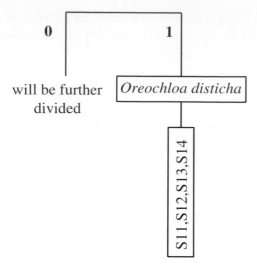

Figure 7-12. The first division in the TWINSPAN example.

We can now start with drawing the dendrogram. The part that is now clear is displayed in Figure 7-12.

The positive group is terminal, because it is smaller than five samples, which is the selected minimum size for division. The next level follows (without the preferentials shown):

```
DIVISION    2  (N= 10)        I.E. GROUP *0
   Eigenvalue  .344 at iteration  1
   INDICATORS, together with their SIGN
   Pice Abie1(-)
Maximum indicator score for negative group   -1    Minimum indicator score for
positive group   0


   Items in NEGATIVE group   4  (N=  7)         i.e. group *00
   Samp0001  Samp0002  Samp0003  Samp0004  Samp0005  Samp0006  Samp0007


   Items in POSITIVE group   5  (N=  3)         i.e. group *01
   Samp0008  Samp0009  Samp0010


DIVISION    3  (N=  4)        I.E. GROUP *1
   DIVISION FAILS - There are too few items
     End of level    2
```

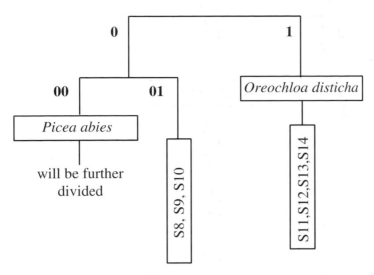

Figure 7-13. The second-level division in the TWINSPAN example.

In a similar way, we can continue to construct the dendrogram (Figure 7-13). The next division level is displayed here:

```
DIVISION     4  (N= 7)              I.E. GROUP *00
   Eigenvalue  .279 at iteration   1
   INDICATORS, together with their SIGN
   Pinu Mugo1(+)
Maximum indicator score for negative group   0     Minimum indicator score for
positive group  1

   Items in NEGATIVE group 8  (N=   5)          i.e. group *000
   Samp0001  Samp0002  Samp0003  Samp0004  Samp0005

   Items in POSITIVE group 9  (N=   2)          i.e. group *001
   Samp0006  Samp0007
```

Note the meaning of indicator species *Pinus mugo* here. It is an indicator within group 00 (containing seven samples), where it is present in just two of the samples, 6 and 7 (forming the positive group 001). However, it is also common in samples 8, 9, 10, 11 and 12. This illustrates that the indicators are determined and should be interpreted just in a particular division context, not for the whole data set. Group 001 is characterized by the presence of *Pinus mugo* in contrast to group 000, and not within the whole data set. The advantage of TWINSPAN becomes clear here, as the orientation of groups (i.e. which group will be negative and which positive) depends on which of the groups is more

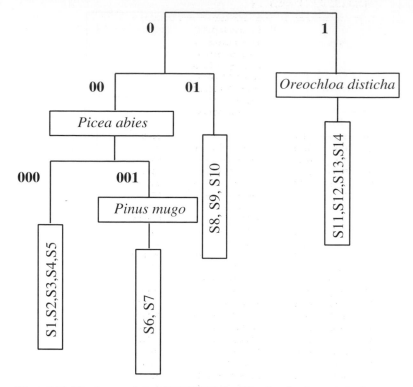

Figure 7-14. Final state of the TWINSPAN classification for the sample data.

similar to group 01 (and this group, among others, has *Pinus mugo* in all of its samples).

We will not show further divisions (group 000 contains five samples and still can be divided further), and will finish the dendrogram at this level (Figure 7-14). The fact that each division is accompanied by only one indicator is rather exceptional (probably a consequence of the small data size). Usually, each division is accompanied by several indicators for each part of the dichotomy. However, note that the indicators are the species used in the divisions. They might be very few and might be available only for one side of the dichotomy. This does not mean that there are no more species characteristic of either side of the dichotomy. If you want to characterize the division by more species, you should use the 'preferentials' from the TWINSPAN output.

In a similar way, TWINSPAN also constructs the classification (dendrogram) for species.

Finally, the TWINSPAN output contains the sorted data table. It resembles a classical ordered phytosociological table and is accompanied by the classification of both samples and species (Table 7-1).

Table 7-1. *Final sorted species data table, produced by TWINSPAN*

```
                        SSSSSSSSSSSSSS
                        aaaaaaaaaaaaaa
                        mmmmmmmmmmmmmm
                        PPPPPPPPPPPPPP
                        00000000000000
                        00000000000000
                        00000000011111
                        21345678901234

 4   Sali   SIle       ---------5----      0000
29   Anth   Alpi       -----2---3----      0000
30   Hype   Macu       ------2--3----      0000
31   Rubu   Idae       ------2--3----      0000
28   Aden   Alli       -----2---2----      0001
 1   Pice   Abie       6666665---5---      001000
 7   Oxal   Acet       55344-4--3----      001001
 9   Sold   Hung       43444---------      001001
18   Luzu   Pilo       2-2-----------      001001
20   Luzu   Sylv       --3243--------      001001
12   Gent   Ascl       23333333------      001010
14   Pheg   Dryo       4333-33-------      001010
15   Pren   Purp       2----3--------      001011
16   Poly   Vert       3----33-------      001011
22   Stel   Nemo       ---2--3-------      001011
 2   Sorb   Aucu       42-23444------      00110
10   Aven   Flex       34543443343---      00110
13   Dryo   Dila       333333-3-3----      00110
17   SoAu   cJu        3-----32------      00110
19   Athy   Dist       3-23335--53---      00111
 6   Vacc   Myrt       54646666636653      01
 8   Homo   Alpi       44454454334334      01
11   Cala   Vill       33365-54-6445-      01
21   Rume   Arif       --23--4--33---      01
 3   Pinu   Mugo       -----3666665--      10
23   Vera   Lobe       ----2333-4322-      10
27   Hupe   Sela       -----223--22-3      10
36   Soli   Virg       ---------2--2-      10
33   Vacc   Viti       -------33-3343      1100
35   Gent   Punc       -------3-4333-      1100
37   Luzu   Luzu       ---------3--4-      1100
24   Ligu   Mute       ----233--23333      1101
25   Sold   Carp       -----54---3333      1101
26   Ranu   Mont       -----2-----33-      1101
32   Geun   Mont       ------2--3-33-      1101
 5   Juni   Comm       -----------24-      111
34   Puls   Alba       -----------32-      111
38   Oreo   Dist       ----------5564      111
39   Fest   Supi       ----------3444      111
40   Camp   Alpi       ----------34-4      111
41   Junc   Trif       ----------4453      111
42   Luzu   Alpi       ----------33--      111
43   Hier   Alpi       ----------233-      111
44   Care   Semp       -----------545      111
45   Tris   Fusc       -----------33-      111
46   Pote   Aure       ------------32      111
47   Sale   Herb       -------------5      111
48   Prim   Mini       -------------4      111

                        00000000001111
                        0000000111
                        0000011
```

Note that we can read the three levels of division and memberships of all the samples in corresponding groups from the bottom three lines. For example, samples 11–14 are in group 1 (which is not further divided).

The TWINSPAN program was written in the late seventies (Hill, 1979). Computer time was precious then and, consequently, very lax convergence criteria were used, which might lead to unstable solutions in some data sets (Oksanen & Minchin, 1997). Be sure that you are using a more recent program version with strict convergence criteria.

8

Regression methods

Regression models allow us to describe the dependence of (usually) one response variable (which might be quantitative or qualitative) on one or more predictor variables. These can be quantitative variables and/or factors. In this broad view, regression models also include statistical methods such as analysis of variance (ANOVA) and the analysis of contingency tables.

During the eighties, many new types of regression models were suggested, usually extending in some way the well-established ones. This chapter provides a short summary of those methods that we find useful for *a posteriori* analysis of the patterns of individual response variables (species) or predictors (environmental variables) in ordination space, or generally for exploration of ecological data.

Most of the regression models introduced in this chapter (except the regression and classification trees) are available in the CanoDraw 4 program and we illustrate their use in the concluding, tutorial, part of this chapter.

8.1. Regression models in general

All regression models share some assumptions about the response variable(s) and the predictor(s). To introduce these concepts, we will restrict our attention to the most frequently used kind of regression model, where exactly one response variable is modelled using one or more predictors.

The simplest way to describe any such type of regression model is the following:

$$Y = EY + e$$

where Y refers to the values of response variable, EY is the value of the response variable expected for particular values of the predictor(s) and e is the variability of the true values around those expected values EY. The expected value of the

Table 8-1. *Summary of the differences between systematic and stochastic components of a statistical model*

Systematic component (EY)	Stochastic component (e)
Is determined by our research hypothesis	Mirrors *a priori* assumptions of the model
Its parameters (regression coefficients) are estimated by fitting the model	Its parameter(s) are estimated during or after fitting (variance of response variable)
We interpret it and perform (partial) tests on it or on its individual parameters	We use it to estimate the model quality (**regression diagnostics**)

response can be formally described as a function of the predictor values:

$$EY = f\left(X_1, \ldots, X_p\right)$$

In the terms of Section 5.1, the EY part is often called the **systematic component** of the regression model, while the e part is called the **stochastic component** of the model. The properties and roles they have, when we apply regression models to our data, are compared in Table 8-1.

When we fit a regression model to our data, our assumptions about its stochastic component* are fixed, but we can manipulate the contents and complexity of the systematic component. In the simplest example of the classical linear regression model with one response variable Y and one predictor X, the systematic component can be specified as:

$$EY = f\left(X\right) = \beta_0 + \beta_1 \cdot X$$

But we can, in fact, achieve a larger complexity in modelling the dependence of Y on X by a polynomial model. We can shift the model complexity along a range starting with a **null model** $EY = \beta_0$, through the linear dependency given above, to the quadratic polynomial dependency $EY = \beta_0 + \beta_1 \cdot X + \beta_2 \cdot X^2$, up to a polynomial of the nth degree, where n is the number of data observations we have available, minus one. This most complex polynomial model (representing the so-called **full model**) goes exactly through all the data points, but provides no simplification of the reality (which is one of the basic tasks of a statistical model). We have simply replaced the n data points with n regression parameters $(\beta_0, \ldots, \beta_{n-1})$. The null model, on the other hand, simplifies the reality so much that we do not learn anything new from such a model,[†] and advance of our knowledge is another essential service provided by statistical models.

---

* Usually assumptions about the distributional properties, the independence of individual observations, or a particular type of cross-dependence between the individual observations.
† Not a completely correct statement: from a null model we learn about the average value of the response variable.

From our discussion of the two extreme cases of regression model complexity, we can clearly see that the selection of model complexity spans a gradient from simple, not so precise models to (overly) complex models. Also, the more complicated models have another undesired property: they are too well fitted to our data sample, but they provide biased prediction for the non-sampled part of the statistical population. Our task is, in general terms, to find a compromise, a model as simple as possible for it to be useful, but neither more simple nor more complex than that. Such a model is often referred to as a **parsimonious** model or the minimal adequate model.

8.2.   General linear model: terms

The first important stop on our tour over the families of modern regression methods is the general linear model. Note the word **general** – another type of regression method uses the word **generalized** in the same position, but meaning something different. In fact, the generalized linear model (GLM), discussed in the next section, is based on the general linear model (discussed here) and represents its generalization.[*]

What makes the general linear model different from the traditional linear regression model is, from the user's point of view, mainly that both the quantitative variables and qualitative variables (factors) can be used as predictors. Therefore, the methods of analysis of variance (ANOVA) belong to the family of general linear models. For simplicity, we can imagine that any such factor is replaced (encoded) by $k - 1$ 'dummy', 0/1 variables, if the factor has k different levels.

In this way, we can represent the general linear model by the following equation:

$$Y_i = \beta_0 + \sum_{j=1}^{p} \beta_j \cdot X_{ji} + \varepsilon$$

but we must realize that a factor is usually represented by more than one predictor X_j and, therefore, by more than one regression coefficient. The symbol ε refers to the random, stochastic variable representing the stochastic component of the regression model. In the context of general linear models, this variable is most often assumed to have a zero mean and a constant variance.

This model description immediately shows one very important property of the general linear model – it is **additive**. The effects of the individual predictors

---

[*] We mean this sentence seriously, really ☺

are mutually independent.* If we increase, for example, the value of one of the predictors by one unit, this has a constant effect (expressed by the value of the regression coefficient corresponding to that variable), independent of the values the other variables have and even independent of the original value of the variable we are incrementing.

The above-given equation refers to the (theoretical) population of all possible observations, which we sample when collecting our actual data set. Based on such a finite sample, we estimate the true values of the regression coefficients β_j, and these estimates are usually labelled as b_j. Estimation of the values of regression coefficients is what we usually mean when we refer to **fitting** a model. When we take the observed values of the predictors, we can calculate the **fitted** (predicted) values of the response variable as:

$$\hat{Y} = b_0 + \sum_{j=1}^{p} b_j \cdot X_j$$

The fitted values allow us to estimate the realizations of the random variable representing the stochastic component – such a realization is called the **regression residual** and labelled as e_i:

$$e_i = Y_i - \hat{Y}_i$$

Therefore, the residual is the difference between the observed value of the response variable and the corresponding value predicted by the fitted regression model.

The variability in the values of the response variable can be expressed by the **total sum of squares**, defined as

$$\text{TSS} = \sum_{i=1}^{n} (Y_i - \overline{Y})^2$$

where $\overline{Y}$ is the mean of Y.

From the point of view of a fitted regression model, this amount of variability can be further divided into two parts – the variability of the response variable explained by the fitted model – the **model sum of squares,** defined as

$$\text{MSS} = \sum_{i=1}^{n} (\hat{Y}_i - \overline{Y})^2$$

---

* This does not mean that the predictors cannot be correlated!

and the **residual sum of squares** defined as

$$\text{RSS} = \sum_{i=1}^{n}(Y_i - \hat{Y}_i)^2 = \sum_{i=1}^{n} e_i^2$$

Obviously TSS = MSS + RSS. We can use these statistics to test the significance of the model. Under the global null hypothesis ('the response variable is independent of the predictors') MSS is not different from RSS if both are divided by their respective number of degrees of freedom.[*]

8.3.   Generalized linear models (GLM)

Generalized linear models (McCullagh & Nelder 1989) extend the general linear model in two important ways.

First, the **expected** values of the response variable (*EY*) are not supposed to be always directly equal to the linear combination of the predictor variables. Rather, the scale of the response depends on the scale of the predictors through some simple parametric function called the **link function**:

$$g(EY) = \eta$$

where η is the **linear predictor** and is defined in the same way as the whole systematic component of the **general** linear model, namely as:

$$\eta = \beta_0 + \sum \beta_j X_j$$

The use of the link function has the advantage that it allows us to map values from the whole real-valued scale of the linear predictor (reaching, generally, from $-\infty$ to $+\infty$) into a specific interval making more sense for the response variable (such as non-negative values for counts or values between 0 and 1 for probabilities).

Second, generalized linear models have less specific assumptions about the stochastic component compared to general linear models. The variance needs not be constant but can depend on the expected value of the response variable, *EY*. This mean–variance relation is usually specified through the statistical distribution assumed for the stochastic part (and, therefore, for the response variable). But note that the mean–variance relation (described by the **variance function**, see Table 8-2), not the specific statistical distribution, is the essential property of the model specification.[†]

---

[*] See any statistical textbook for more details, e.g. Sokal & Rohlf (1995).
[†] This is known as the quasi-likelihood approach to generalized linear modelling.

Table 8-2. *Summary of useful combinations of link functions and types of response variable distributions*

Type of variables	Typical link function	Reference distribution	Variance function (mean-variance relation)
Counts (frequency)	Log	Poisson	$V \propto EY$
Probability (relative frequency)	Logit or probit	Binomial	$V \propto EY \cdot (1 - EY)$
Dimensions, ratios	Inverse or log	Gamma	$V \propto EY^2$
Quite rare type of measurements	Identity	Gaussian ('normal')	$V = \text{const}$

The assumed relation between the variance (V) of the stochastic part and the expected values of response variable (EY) is also indicated.

The options we have for the link functions and for the assumed type of distribution of the response variable cannot be combined independently, however. For example, the **logit** link function maps the real scale onto a range from 0 to $+1$, so it is not a good link function for, say, an assumed Poisson distribution, and it is useful mostly for modelling probability as a parameter of the binomial distribution. Table 8-2 lists some typical combinations of the assumed link functions and the expected distribution of the response variable, together with a short characteristic for the response variables matching these assumptions. Note that as with classical linear regression models and ANOVA, it is not assumed that the response variable has the particular types of distribution, but that it can be **reasonably approximated** by such a distribution.

With this knowledge, we can summarize what kinds of regression models are embraced by the GLMs:

- 'classical' general linear models (including most types of analysis of variance)
- extension of those classical linear models to variables with non-constant variance (counts, relative frequencies)
- analysis of contingency tables (using log-linear models)
- models of survival probabilities used in toxicology (probit analysis)

The generalized linear models extend the concept of a residual sum of squares. The extent of discrepancy between the true values of the response variable and those predicted by the model is expressed by the model's **deviance**. Therefore, to assess the quality of a model, we use statistical tests based on the **analysis of deviance**, quite similar in concept to an analysis of variance of a regression model.

An important property of the general linear model, namely its **linearity**, is retained in the generalized linear models on the scale of the linear predictor. The effect of a particular predictor is expressed by a single parameter – linear transformation coefficient (the regression coefficient). Similarly, the model additivity is kept on the linear predictor scale. On the scale of the response variable, things might look differently, however. For example with a logarithmic link function, the additivity on the scale of linear predictor corresponds to a multiplicative effect on the scale of the response variable.

8.4.   Loess smoother

The term **smoother** is used for the method of deriving a (non-parametric) regression function from observations. The fitted values produced by smoothing (i.e. by application of a smoother) are less variable than the observed ones (hence the name 'smoother').

There are several types of smoothers, some of them not very good, but simple to understand. An example of such a smoother is the moving average smoother. An example of a better smoother is the **loess smoother** (earlier also named *lowess*). This smoother is based on a locally weighted linear regression (Cleveland & Devlin 1988; Hastie & Tibshirani 1990). The area (band for a single predictor) around the estimation point, which is used to select data for the local regression model fit, is called the **bandwidth** and it is specified as a fraction of the total available data set. Therefore, bandwidth value $\alpha = 0.5$ specifies that at each estimation point half of the observations (those closest to the considered combination of the predictors' values) are used in the regression. The complexity of the local linear regression model is specified by the second parameter of the loess smoother, called **degree** (λ). Typically, only two values are available: 1 for a linear regression model and 2 for a second-order polynomial model.

Furthermore, the data points used to fit the local regression model do not enter it with the same weight. Their weights depend on their distance from the considered estimation point in the predictors' space. If a data point has exactly the required values of the predictors, its weight is equal to 1.0 and the weight smoothly decreases to 0.0 at the edge of the smoother bandwidth.

An important feature of the loess regression model is that we can express its complexity using the same kind of units as in traditional linear regression models – the number of degrees of freedom (DF) taken from the data by the fitted model. These are, alternatively, called the **equivalent number of parameters**. Further, because the loess model produces fitted values of the response variable (like other models), we can work out the variability in the values of the response accounted for by the fitted model and compare it with the residual sum of squares. As we have the number of DFs of the model estimated, we can calculate the residual DFs and calculate the sum of squares per one degree of freedom (corresponding to the mean square in an analysis of variance of a classical regression model). Consequently, we can compare loess models using an analysis of variance in the same way we do for general linear models.

Good alternative smoothers are the various types of smoothing splines, see Eubank (1988) for more details.

8.5.   Generalized additive models (GAM)

The generalized additive models (GAMs, Hastie & Tibshirani 1990) provide an interesting extension to generalized linear models (GLMs). The so-called **additive predictor** replaces the linear predictor of a GLM here. It is also represented by a sum of independent contributions of the individual predictors; nevertheless the effect of a particular predictor variable is not summarized using a simple regression coefficient. Instead, it is expressed – for the jth predictor variable – by a smooth function s_j, describing the transformation from the predictor values to the (additive) effect of that predictor upon the expected values of the response variable.

$$\eta_A = \beta_0 + \sum s_j(X_j)$$

The additive predictor scale is again related to the scale of the response variable via the link function.

We can see that generalized additive models include generalized linear models as a special case, where for each of the predictors the transformation function is defined as:

$$s_j(X_j) = \beta_j \cdot X_j$$

In the more general case, however, smooth transformation functions (usually called 'smooth terms') are fitted using a loess smoother or a cubic spline smoother. When fitting a generalized additive model, we do not prescribe the shape of the smooth functions of the individual predictors, but we must specify the complexity of the individual curves, in terms of their degrees of freedom. We also need to select the type of smoother used to find the shape of the smooth transformation functions for the individual predictors.

With generalized additive models, we can do a stepwise model selection including not only a selection of the predictors used in the systematic part of the model but also a selection of complexity in their smooth terms.

Generalized additive models cannot be easily summarized numerically, in contrast to generalized linear models where their primary parameters – the regression coefficients – summarize the shape of the regression model. The fitted additive model is best summarized by plotting the estimated smooth terms representing the relation between the values of a predictor and its effect on the modelled response variable.

8.6.   Classification and regression trees

Tree-based models are probably the most non-parametric kind of regression models one can use to describe the dependence of the response variable values on the values of the predictor variables. They are defined by a recursive binary partitioning of the data set into subgroups that are successively more and more homogeneous in the values of the response variable.* At each partitioning step, exactly one of the predictors is used to define the binary split. The split that maximizes the homogeneity and the difference between the resulting two subgroups is then selected. Each split uses exactly one of the predictors – these might be either quantitative or qualitative variables.

The response variable is either quantitative (in the case of **regression trees**) or qualitative (for **classification trees**). The results of the fitting are described by a 'tree' portraying the successive splits. Each branching is described by a specific splitting rule: this rule has the form of an inequality for quantitative predictors (e.g. 'VariableX2 $<$ *const*') or the form of an enumeration of the possible values for a factor (e.g. 'VariableX4 has value *a, c,* or *d*'). The two subgroups defined by such a split rule are further subdivided, until they are too small or sufficiently homogeneous in the values of the response. The terminal groups (leaves) are then identified by a predicted value of the response value (if this is a quantitative variable) or by a prediction of the object membership in a class (if the response variable is a factor).

When we fit a tree-based model to a data set, we typically create ('grow') an overcomplicated tree based on our data. Then we try to find an optimum size for the tree for the prediction of the response values. A **cross-validation procedure** is used to determine the 'optimum' size of the tree. In this procedure, we create a series of progressively reduced ('pruned') trees using only a subset of the available data and then we use the remaining part of the data set to assess the performance of the created tree: we pass these observations through

---

* The division is very similar to recursive binary splitting done by TWINSPAN, but there the multivariate data set is split and also the splitting method is different.

the hierarchical system of the splitting rules and compare the predicted value of the response variable with its known value. For each size ('complexity') of the tree model, we do it several times. Typically, we split the data set into 10 parts of approximately the same size and, for each of these parts, we fit a tree model of given complexity using the remaining nine parts and then we use the current one for performance assessment. A graph of the dependency of the tree model 'quality' on its complexity (model size) has typically a minimum corresponding to the optimum model size. If we use a larger tree, the model is overfitted – it provides a close approximation of the collected sample, but a biased description of the sampled population.

8.7.   Modelling species response curves with CanoDraw

This section is a short tutorial for fitting the various regression models that describe the relationship between a quantity of a particular species and the environmental gradients or gradients of community variation. The process of fitting various regression models (generalized linear models, generalized additive models and loess smoother) will be illustrated for the CanoDraw for Windows program, distributed with CANOCO. In this tutorial, you will study the change in abundance of six selected sedge species in wet meadow communities, along the compositional gradient represented by the first ordination axis of DCA. This sample study uses the same data set that is used in Case study 2 (Chapter 12), and you should check that chapter for a more detailed description. The selected sedge species belong to the more frequently occurring sedges in the species data and their optima (approximated by their scores in the ordination space) span the whole range of compositional variation in the sampled community types, as can be seen from Figure 8-1.

The first ordination axis represents a gradient of increasing trophic level of underground water (from left to right in the diagram) and decreasing concentration of Ca, Mg and Na ions. In your regression models describing the change of species abundances, the 'species response curves', you will use the scores of individual samples on the first DCA axis as the explanatory variable. The results of the detrended correspondence analysis are stored in the file containing CanoDraw project, *regress.cdw*. Open the project in CanoDraw for Windows if you want to work through this tutorial.

Start with building a regression model for *Carex panicea* (labelled as *CarxPani* in the sample project). To get your first impression about the change of abundance of this species along the first ordination axis, create an *XY*-scatterplot and supplement it with the non-parametric loess model. To do so, use the *Create > Attribute Plots > XY(Z) Plot* and in the dialog box select *Analysis Results > Sample*

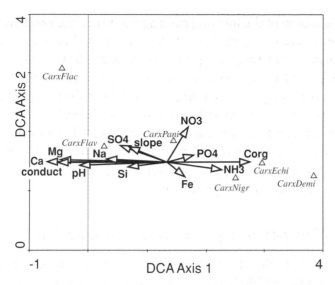

Figure 8-1. Biplot of environmental variables and the selected six *Carex* species, based on a DCA method.

scores > Samp.1 in the *X VARIABLE* list. In the *Y VARIABLE* list select *Data Files > Species data* and there the *CarxPani* item. In the lower left corner select the *Loess* option and also check the *Supplement model with datapoints* option. The desired final state of the dialog is illustrated in Figure 8-2.

Often, you will be interested in a species response to measured environmental variables, rather than to compositional gradients (represented by the ordination axes). In such a case, you should select in the *X VARIABLE* area the required variable in one of the folders within the *Data Files* or the *Imported data* folders.

After you press the *OK* button, you must confirm or change the options for the loess model. Keep the default choices there (*Local linear model*, no *Parametric fit*, no *Drop square*, *Normalize scale*, and *Robust fitting*, and *Span* equal to 0.67). After the model is fitted, CanoDraw summarizes it in another dialog. Close the dialog by clicking the *OK* button. CanoDraw displays the diagram, similar to the one shown in Figure 8-3.

Note that in your diagram the individual data points are labelled. To remove the labels, select any one of them by clicking it with the left mouse button and then press *Ctrl-H* to select the other labels. Remove the labels by pressing the *Delete* key on the keyboard. You can see in the diagram that the vertical coordinates of plotted points are surprisingly regularly spaced. This is because the abundances of individual species in this data set were recorded on a semi-quantitative estimation scale with values ranging from 0 to 7.

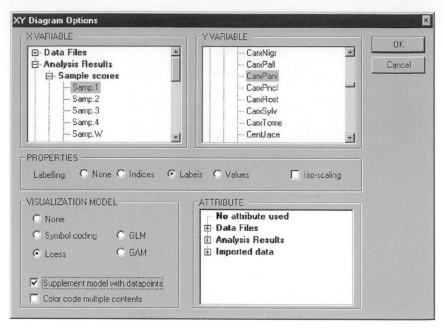

Figure 8-2. Selections to be made in the *XY* Diagram Options dialog when fitting the loess model.

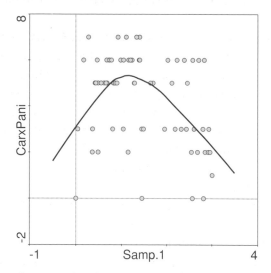

Figure 8-3. Abundance of *Carex panicea* plotted against the first DCA axis together with a fitted loess model.

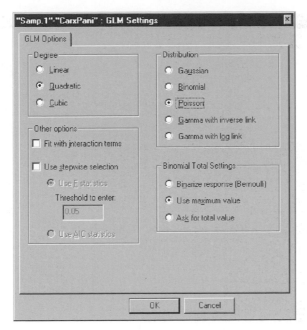

Figure 8-4. GLM Settings dialog: options for fitting unimodal response curves.

The curve estimated by the loess smoother suggests a unimodal shape of the response.

You will now fit the classical unimodal response curve ('Gaussian model', see Section 3.9.3 in Ter Braak & Šmilauer 2002) to the same data, i.e. response curve describing change of *CarxPani* values along the first ordination axis. Repeat the steps used in the previous fitting, including the selection of variables in the *XY Diagram Options*, as illustrated in Figure 8-2. Change in this dialog box the following options: use the *GLM* choice instead of *Loess* in the *VISUALIZATION MODEL* area and in the middle part of the dialog change the *Labelling* option value from *Labels* to *None* (to remove sample labels more easily than for the previous diagram).

After you press the *OK* button, CanoDraw displays new dialog box where you can modify the specification for the fitted generalized linear model. The required settings are illustrated in Figure 8-4.

You are fitting a generalized linear model with the predictor variable (scores of samples on the first ordination axis) used in the form of a second-order polynomial. In addition, you specify the *Poisson* distribution, with an implied *log* link function. Note that we **do not** assume that the response variable values really have the Poisson distribution. But we find this is the best choice amongst the available options, particularly because the (implied) log link function is

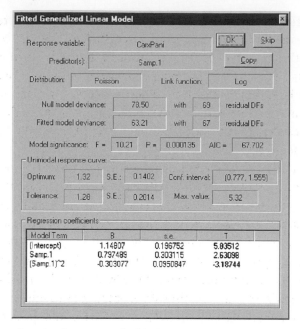

Figure 8-5. Summary of a fitted unimodal response curve.

needed to fit the Gaussian-shaped response curve (with the tails approaching zero), rather than the plain second-order polynomial parabolic curve. Additionally, the choice of Poisson distribution implies that the variance is proportional to the mean (see Table 8-2), and this does not seem unreasonable. After you close this dialog with the *OK* button, the regression model is fitted and fitting results are summarized in another dialog box, shown in Figure 8-5.

Information in the upper part of the dialog summarizes the model properties, which you specified in the preceding dialogs. The fitted model is compared with the null model ($EY = $ const) and the dialog shows both the raw deviance values for both models as well as a deviance-based F test comparing the difference between the deviances of the null model and the fitted model (in the numerator of the F statistic ratio) with the deviance of the null model (in the denominator). Both terms are adjusted for the number of degrees of freedom, so they correspond to the mean-square terms in the traditional F statistic used in ANOVA models.

In the white area at the bottom of this summary dialog, the estimated values of individual regression coefficients, as well as the standard errors of the estimates are shown.* The last column gives the ratio between the estimate and its

---

* CanoDraw ignores the estimated scale parameter (or dispersion parameter) in these error estimates.

standard error. In classical regression models these ratios are used for testing the partial hypotheses about individual regression coefficients, i.e. $H_0 : b_j = 0$. Under the null hypothesis (and fulfilled assumptions about the distributional properties of the response variable and independence of individual observations), these statistics have the t (Student) distribution. Note, however, that this approximation does not work very well for the whole family of generalized linear models, so comparison of the values with the critical values of the t distribution is not recommended.

The area above the list of coefficient estimates is used only if a second-order polynomial model was fitted, using either the *log* or *logit* link function (i.e. the *Distribution* type was specified as *Binomial*, *Poisson*, or *Gamma with log link*). In such cases, CanoDraw attempts to estimate the two parameters of the species response curve. The first parameter is the position of the curve maximum on the gradient represented by the predictor variable (*Samp.1* in your case), called *Optimum*. The second parameter is the 'width' of the fitted curve, named *Tolerance*. Depending on the meaning of the predictor variable, the tolerance parameter can be sometimes interpreted as the width of the species niche with respect to a particular resource gradient ('niche dimension'). Depending on the quality of the fitted unimodal model, CanoDraw can provide the estimated variability of the two parameters (the *S.E.* fields) and, in the case of the optimum, estimate the 0.95 confidence interval for the estimate. The *Max. value* field specifies the fitted (predicted) abundance of the species at the predictor value equal to the optimum estimate (i.e. the 'height of the curve').

The estimated values of the tolerances and optima of individual species can be used for further data exploration or testing of hypotheses concerning the behaviour of individual species. Note, however, that you must be careful to select an appropriate predictor for such models. If you use the sample scores on the first DCA axis as the predictor variable, you can use the resulting models to illustrate in much detail the meaning of the ordination results, but you cannot draw any independent conclusions from the positions of species optima (or change in species tolerances) along the DCA axis. This is because the dispersion of the species scores, as well as the variation of the sample scores (affecting the width of the fitted response curves) are systematically modified by the weighted averaging (and also the detrending) algorithm.

To obtain valid tests of differences among the species in relation to resource gradients, you would need to specify in CanoDraw independently measured explanatory variables as your predictors. In that case, the ordination analysis performed with CANOCO is not involved at all, but it still provides a useful 'framework', representing a rich source of ideas about which environmental

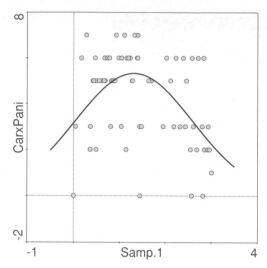

Figure 8-6. Fitted species unimodal response model, using GLM with a second-order polynomial predictor.

gradients can be important for variation in community composition and which species respond to which factors.

The fitted regression model is shown in Figure 8-6.

The second-order polynomial is a very strict specification for the shape of the response curve. The response of species often takes more complicated shapes and it is difficult to describe them with the more complicated polynomial terms. Therefore, other families of regression models can be useful here, e.g. generalized additive models. Before you try to fit the response curve for *CarxPani* with a generalized additive model, you should first explore whether the just-fitted regression model provides an adequate description of the species behaviour along the DCA axis.

To do so, we will plot the GLM regression residuals against the predictor values (*Samp.1*). Click onto any empty place in the diagram with the right mouse button and select the *Residual plots* command from the pop-up menu. CanoDraw displays a dialog where you can specify what to plot in the regression diagnostics plot and how to plot it. Specify the options as illustrated in Figure 8-7. You select the *Raw residuals* option rather than the *Absolute values* one because this type of residual is better for identifying an inappropriate shape of the fitted response curve along the ordination axis. The absolute residual values (eventually square-root transformed) are better for identifying change in the variability with the fitted response values (termed heteroscedasticity in classical regression models). You also checked the *Add LOESS model* option because the smoother line can help you to identify any such systematic

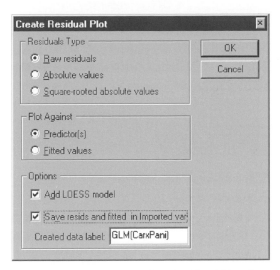

Figure 8-7. Residual plot settings for identifying the appropriateness of model type and complexity.

deviation from the postulated polynomial model. Note that you must also ask to save the residuals and fitted values here. You will use these data when comparing two types of fitted response curve later in this tutorial.

Before the residual plot is shown, CanoDraw asks you to approve or change settings for the loess smoother used to summarize behaviour of the regression residuals with respect to predictor values (keep the default settings there) and then summarizes the fitted loess model. Note that the predictor variable does not explain much about the residual variability (coefficient of determination, labelled *Multiple-R squared*, has a value of 0.027, implying that less than 3% of the variability was explained). The resulting graph is shown in Figure 8-8. In our opinion, there is no substantial trend neglected by the unimodal generalized linear model. The regularly curved patterns of the points are a consequence of the limitation of the response variable values to whole numbers.

You will now compare the fitted response curve with another one, estimated with a generalized additive model (GAM). To compare the two kinds of models 'head-to-head', you need to fit a generalized additive model not only with the same type of distribution assumed for the response variable (and with the same link function), but also with a comparable amount of complexity. To compare with a second-order polynomial model, you must specify smooth term complexity as 2 DF. But we think it is better to let CanoDraw compare models of differing complexity (but with an identical predictor, i.e. the sample scores on the first DCA axis) and select the best one. CanoDraw will measure the performance

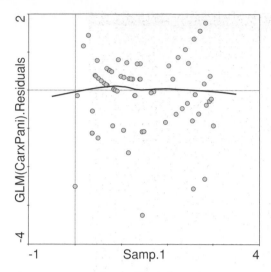

Figure 8-8. Regression residuals from the GLM model plotted against the predictor values.

of individual candidate models using the *Akaike Information Criterion* (AIC). This criterion attempts to measure model 'parsimony', so it penalizes a candidate model for its complexity, measured using the number of model degrees of freedom (see Hastie & Tibshirani 1990, or Chambers & Hastie 1992, for additional discussion).

So, start your fitting with the same dialog box (XY Diagram Options), but specify the *GAM* option in the lower left corner of this dialog. When you close it, CanoDraw displays dialog where you set options for the fitted generalized additive model. Make sure you change the default settings to match those shown in Figure 8-9.

You specified value 4 in the *Predictor 1* ($DF=$) field. If the *Stepwise selection using AIC* option was not selected, a generalized additive model with a smooth term for *Samp.1* with complexity of four degrees of freedom would be fitted. But the model selection option is enabled (checked), so CanoDraw will fit a null model (which is, in fact, identical to the null model we mentioned when fitting GLM) and four alternative generalized additive models with increasing complexity (*DF* equal to 1, 2, 3, 4). If you specify a non-integer number in the *Predictor 1* field, CanoDraw fits a model with the integer sequence of DFs up to the integer immediately below the specified value and then the model with the specified complexity. For example, if the field value were 3.5, CanoDraw would make a comparison between a null model and the models with 1, 2, 3, and 3.5 DFs. Note that the distribution specification is identical to the one you used when fitting the generalized linear model.

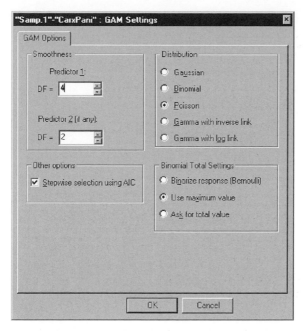

Figure 8-9. Settings for generalized additive model with model complexity
selection.

After you press the *OK* button, CanoDraw selects the model complexity and
informs you about it with the *Model Selection Report* dialog. We reproduce its contents here:

```
AIC-BASED MODEL SELECTION
Response: CarxPani
Predictor: Samp.1
Selected model is marked by asterisk
  Model            AIC
------------------------
Null model        80.16
  s(X,1)          77.19
  s(X,2)          69.18
* s(X,3)          68.66
  s(X,4)          69.14
```

As you can see, the model with three degrees of freedom (s(X, 3)) has the lowest AIC value (the highest parsimony). This is the one which CanoDraw accepts,
summarizes in the following dialog box, and plots in the diagram illustrated
in Figure 8-10.

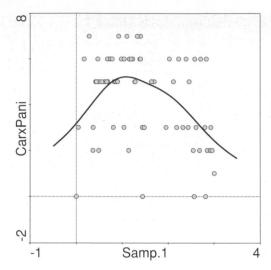

Figure 8-10. Species response curve fitted with a generalized additive model with df = 3.

The curve has an asymmetrical shape, but in general corresponds well to the generalized linear model you fitted before. To compare these two models together, you can plot fitted values of *CarxPani* from one model against the fitted values from the other model. Alternatively, you can similarly plot the residuals from the two models. The two alternatives provide complementary information, because the residuals are defined as the actual response variable values minus the fitted values, and the two models share the values of the response variable. Whatever option we choose, we need to store the fitted values and residuals of this generalized additive model in the project and there is no other way to do so than to create a residual plot. So, repeat the procedure illustrated for the generalized linear model and make sure that the *Save resids and fitted* option is checked. Note that no difference from the previously created residual plot can be seen in the new residual plot (not shown here).

You will compare the fitted GAM and GLM by plotting the residuals from GAM against the residuals from GLM. To do so, use again the *Create > Attribute Plots > XY(Z) Plot* command, and make your choices as shown in Figure 8-11.

Note that no regression model is fitted this time (option *None* in the *VISUALIZATION MODEL* field) and that the diagram aspect ratio is fixed with the *Isoscaling* option. The resulting graph is shown in Figure 8-12.

You can see there is little discrepancy between the two fitted models.

Finally, we illustrate how to fit and plot species response models for multiple species at the same time, determining the appropriate model complexity for each species (response variable) independently. You should return to the

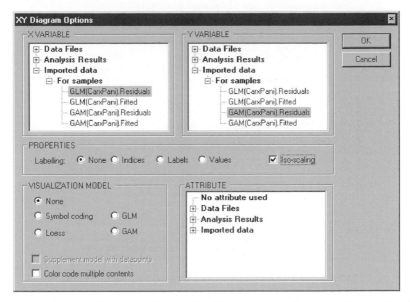

Figure 8-11. XY Diagram Options for the diagram comparing the residuals of two regression models.

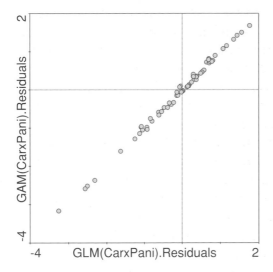

Figure 8-12. XY diagram comparing the residuals of two alternative regression models.

beginning of this section and study the diagram in Figure 8-1, containing six selected sedge species which you will compare with respect to their distribution along the main gradient of meadow community variation, represented by the first DCA axis. Because it is quite possible that you may need to refer to this species group in several diagrams, you should define the group explicitly

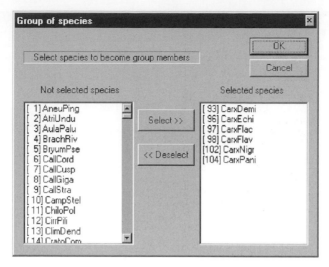

Figure 8-13. Definition of the group of sedge species.

within the project. Go to the *Project* menu and select the *Define Groups of >
Species*. A new dialog (named *Species Group Manager*) appears and you should
click the *By Selection* button in the *Create* area. Another dialog appears and you
should select the appropriate species names in the left list and move them to
the right list by clicking the *Select* button. The desired final state is illustrated
in Figure 8-13.

After you close this dialog with the *OK* button, you can change the default
group name (*Group of species 1*) using the *Rename* button in the group manager
dialog. Leave the dialog using the *Close* button.

To create a diagram with response curves of multiple species, you can use
the same command you used when you fitted response curves for *CarxPani*,
because you can select multiple variables in the *Y VARIABLE* list. But you will
take a shortcut here and use the *Create > Attribute Plots > Species response curves*
command. In the dialog box, which appears after you select the command,
select the *Generalized Additive Model (GAM)* option in the *Response Model Type* area.
In the middle part, select the name of the group you just created in the right-
hand list. You can check that the group members were selected in the left-
hand list (titled *Species to plot*). Now you should select the predictor variable.
Click the *Axis 1* in the left listbox in the lower part of the dialog. As you can
see, you can alternatively fit the response curves with respect to individual
environmental variables. If you want to use a different kind of predictors (e.g.
sample richness, values of covariates, or values of supplementary variables),
you must use the XY(Z) diagram command. Confirm your choices with the *OK*
button.

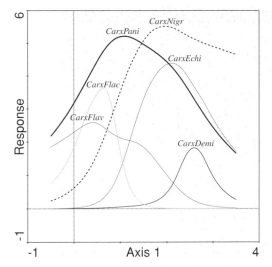

Figure 8-14. Response curves of six sedge species, fitted using generalized additive models.

Note that in the dialog that appears now (titled GAM Options) you will specify options for all the response variables (species) at once. To compare multiple species models, you probably want to keep the fitted curves simple. Specify, therefore, value 3 in the *Predictor 1* field. Use the *Poisson* option in the *Distribution* area and make sure the *Stepwise selection using AIC* option is checked.

After you close this dialog, CanoDraw selects 'optimum' model complexity for each of the specified species and reports about the model selection and about the selected model properties. Finally, the diagram is drawn, similar to the one in Figure 8-14.

Note that your diagram has all the response curves drawn with solid lines of different colour, but with uniform width. We modified the line appearance to better support the black and white reproduction.

We will remind you here that the roughly unimodal shape of the response curves for all the species is not an artefact of the specified model type (unlike the situation where you would fit second-order polynomial GLMs without model complexity selection). Generalized additive models are able to fit monotonically changing or a bimodal response shape, even with the limit of three degrees of freedom imposed in our example.

9

Advanced use of ordination

This chapter introduces four specialized or more advanced techniques, which build on the foundations of ordination methods and can be used with the Canoco for Windows package. All four methods are illustrated in one of the case studies presented in Chapters 11–17.

9.1. Testing the significance of individual constrained ordination axes

If you use several (partially) independent environmental variables in a constrained ordination, the analysis results in several constrained (canonical) ordination axes. In CANOCO, it is easy to test the effect of the first (most important) constrained axis or the effect of the whole set of constrained axes.

But you may be interested in reducing the set of constrained axes (the dimensionality of the canonical ordination space), and to do so you need to find how many canonical axes effectively contribute to the explanation of the response variables (to the explanation of community variation, typically). To do so, you must test the effects of individual constrained axes. Their independent (marginal) effects do not differ from their additional (conditional) effects, of course, because they are mutually linearly independent by their definition.

Let us start with the simplest situation, when you have a constrained ordination (RDA or CCA) with no covariables. If you need to test the significance of a second (or higher) canonical axis for such constrained analysis, you should clone the corresponding CANOCO project, i.e. create a new project, similar to the original one but containing, in addition, covariable data. You will use the scores of the samples on the constrained axes, which were calculated in the original project. If you specify the sample scores (environment-derived scores *SamE*) on the first axis as your only covariable, CANOCO will, in the new analysis, attempt to construct the constrained axes so that they are independent of

140

this covariable. Therefore, the first axis found in the new analysis will be identical to the second constrained axis of the original analysis. Now, you can simply use the test of the first canonical axis, which is directly available in the CANOCO program. This test will, in effect, refer to the original second constrained (second most important) axis. Similarly, if you want to test the effect of a third constrained axis, you must specify both the first and the second constrained axes of the original analysis as two covariables in a new project.

To turn one or more constrained axes into covariables, you can open the solution (.sol) file produced in the original project with the Microsoft Excel program and extract the *SamE* scores from there with the help of the WCanoImp program (see Section 1.7). Note that the constrained sample scores (which are defined as linear combinations of the environmental variables) are usually placed at the very end of the solution file. The sample scores calculated from the species scores (the *Samp* scores) are near the beginning of the file, so make sure you do not use them by mistake.

There is a shortcut, which allows you to skip the export of the sample scores from the solution file. CANOCO is able to use a solution file as its input data. The problem with this approach is, however, that CANOCO reads the incorrect set of sample scores (the *Samp* scores, not the *SamE* scores), unless you modify the solution file manually in a way that causes CANOCO to miss the first occurrence of sample scores. The required change to the solution file is described in Section 8.2.4.2 of the CANOCO reference manual (Ter Braak & Šmilauer 2002).

If your original constrained analysis contained some covariables, you must, of course, combine these covariables with the constrained sample scores, so you must create a new data file before you can analyse the project testing the significance of second and higher constrained axes.

Also note that the above-suggested algorithm for testing the higher constrained axes cannot be used in a DCCA (detrended canonical correspondence analysis). If you use detrending by segments, the tests cannot be performed at all, and if you select detrending by polynomials the required algorithm is much more complicated and it is described in Section 8.2.4.3 of the CANOCO reference manual (Ter Braak & Šmilauer 2002).

Check Section 12.2 in Case study 2 for an example of testing the significance of the second constrained axis.

9.2.   Hierarchical analysis of community variation

If you study variation in a biotic community at several spatial or temporal scales, you must take into account the hierarchical arrangement of the individual scale levels when analysing the data. If your sampling design allows doing so, you may ask questions that involve the hierarchical nature of your data:

1. How large are the fractions of the total community variation that can be explained at the individual scale levels?
2. Is it possible to identify levels that influence significantly the community composition?

In this section, we will illustrate the required procedures on a theoretical level. An example of studying community variation on different spatial (landscape) levels is provided by Case study 6 in Chapter 16, using crayfish communities as an example.

Let us start with the definitions of terms used in the following discussion. We will assume your species data (Y) were collected at three spatial levels. Let us assume a hypothetical project, for which you can imagine that three mountain ranges were sampled, in each of them three mountain ridges were selected (at random), and then three peaks were selected within each of the nine ridges. Species data were collected using five randomly positioned samples on each of the 27 mountains (peaks). We will describe the spatial location of each sample by three dummy variables representing mountain ranges (SLA1 to SLA3), by another nine dummy variables identifying the ridges (SLB1 to SLB9), and another 27 dummy variables identifying the mountain peaks (SLC1 to SLC27). The term 'variation explained by' refers to classical variance when linear ordination methods (PCA, RDA) are used. When weighted-averaging (unimodal) ordination methods (CA, CCA) are used, it refers to so-called **inertia**. For simplicity, we will refer to linear methods when describing the required kind of analysis.

Total variation

The total variation in species (response) data can be calculated using the unconstrained PCA, with no environmental variables or covariables. We can describe the corresponding ordination model (similar to the formal description of regression models, used in some statistical packages) as:

$$Y = \text{const}$$

i.e. this is the null model, with no predictors.

The total variation in X can be decomposed into four additive components: variation explained at the range level, variation explained at the ridge level, variation explained at the mountain peaks level and the within-peak (or residual) variation.

Variation among ranges

To estimate the variation explained at the highest spatial level of ranges (SLA), in an RDA we must use the SLA variables as environmental variables (predictors) and ignore the other SLx variables, i.e. the ordination model will be:

$$Y = SLA$$

To test this effect (the differences among mountain ranges), we can use a Monte Carlo permutation test where we randomly reassign the membership in SLA classes (i.e. which mountain range the data come from). We should keep the other, lower-level spatial arrangement intact (not randomized), however, so that we test only the effect at this spatial level. Therefore, we must permute the membership of whole mountain ridges within the ranges, i.e. we must use a split-plot design where the ridges (**not** the ranges) represent the whole plots and the groups of 15 samples within each ridge represent the split-plots. The whole plots will be permuted randomly (they will randomly change their membership in ranges), while the split-plots will not be permuted at all.

Variation among ridges

When estimating the variation explained at the intermediate spatial level of ridges (SLB), we must use the SLB variables as predictors. This is not enough, however, because their marginal (independent) effect includes the effect at the higher spatial level (SLA), which we have estimated already.[*] Therefore, the SLA variables should be used as covariables in the partial RDA. We can formally describe this partial constrained ordination as:

$$Y = SLB \mid SLA$$

i.e. we extract from X the effect of level SLB, after we have accounted for the effect at level SLA. In other words, we model here the variability **among** ridges, but **within** ranges.

When testing the SLB level effect (differences between the ridges), we must randomly (in an unrestricted way) permute the membership of mountain peaks within the ridges (with all five samples from each peak holding together during the permutations), but we shall not permute across the mountain ranges (the SLA level). To achieve this, we must do the restricted (split-plot type) permutations within blocks defined by the SLA covariables.

---

[*] We have nine ridges: from the identity of a ridge (coded by SLB1 to SLB9 dummy variables), we can tell to which mountain range a sample belongs.

Variation among peaks

The variation explained by the mountain peaks can be estimated (and tested) with a similar (partial RDA) setup as used before, except we are now working at a lower level. We will use the SLC variables as environmental variables and the SLB variables as covariables:

$$Y = \text{SLC} \mid \text{SLB}$$

Further, we will not use the split-plot design permutation restrictions here. Instead, we will randomly allocate mountain peak identity to our samples, but only within the permutation blocks defined by the SLB covariables (i.e. within each of the nine mountain ridges).

Residual variation

We still have one hierarchical component to estimate. It is the variation in community composition among the individual samples collected at the individual peaks. This is the residual, lowest-level variation, so we cannot test it. To estimate its size, we will perform **partial unconstrained** ordination, i.e. a PCA with SLC variables used as the covariables and with no environmental variables.

Note also the following aspect of hierarchical partitioning of community variation. As we move from the top levels of the hierarchy towards the lower levels, the number of degrees of freedom for spatial levels increases exponentially: we have only three mountain ranges, but nine ridges and 27 peaks. Predictably, this fact alone will lead to the increasing size of variation explained by the progressively lower levels. The amount of explained variation (represented by the sum of canonical eigenvalues) corresponds to the sum-of-squares explained by a regression term in a linear regression, or by a factor in an ANOVA model. To compare the relative importance of individual hierarchical levels, we should divide the individual variation fractions by their respective degrees of freedom. The method for calculating the number of DFs for individual hierarchical levels is outlined in Section 16.3.

9.3.   Principal response curves (PRC) method

When we experimentally manipulate whole communities, we often evaluate the effect of the experimental treatments over a longer period. The response to disturbance imposed on an ecosystem or to the addition of nutrients has a strong temporal aspect. When we need to compare the community composition at sites differing in experimental treatment, with the

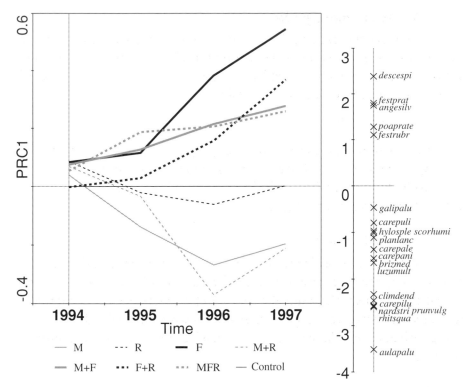

Figure 9-1. Sample diagram with principal response curves, from Section 15.8. M is for mowing the plot; F is for its fertilization and R is for the experimental removal of the dominant plant species.

control (unmodified) plots at different sampling times, it is quite difficult to do so using an ordination diagram from a standard constrained ordination. The temporal trajectory is usually not running straight through the ordination diagram, so it is difficult to evaluate the extent of differences among the individual time steps and also among the individual treatments. Van den Brink & Ter Braak (1998, 1999) developed a new method called principal response curves (PRC), which focuses exactly on this aspect of the data.

The primary result of the PRC method is one or several sets of response curves, representing temporal trajectories of community composition for each of the experimental treatments. An example of a set of eight principal response curves (corresponding to eight experimental treatments applied to grassland vegetation) is shown in Figure 9-1.

A diagram displaying principal response curves can be usefully supplemented by a one-dimensional diagram showing the species scores on the corresponding RDA axis, as illustrated in Figure 9-1. We can combine the value read from a PRC for a particular treatment and time with the species score: if we calculate an exponential transformation of their multiple, the value predicts the relative size of that species abundance in comparison to its abundance in the plots with the control treatment at the same time. For example, the species *Scorzonera humilis* (*scorhumi*) has its score equal to -1.0 on the first RDA axis. We can predict from the diagram that in the third year its cover will be 32% **lower** on the fertilized-only plots, compared with the control plots. This is because the F curve value in 1996 is (approximately) 0.38, so we predict the relative species cover as $\exp(-1.0 \cdot 0.38) = 0.68$. Note that such a quantitative interpretation rule requires log-transformation of the original species data and that it was derived with the assumption of count data. However, the interpretation also carries on to the log-transformed cover estimates used in the study from which Figure 9-1 originates.

The PRC method is based on a partial redundancy analysis. If we have K treatment levels, coded as $Z_1, \ldots, Z_K$ dummy variables in a CANOCO data file, and we measure the community composition at permanent sites at L time points, coded as $T_1, \ldots, T_L$ dummy variables, then we should set up the redundancy analysis model where T variables are used as covariables, and the interactions of the Z and T variables are used as the environmental variables. Because we want the control treatment to represent a reference point for each measurement time, we must omit the interactions involving the Z_1 variable (if the control treatment is represented by this variable).

The canonical (regression) coefficients (the *Regr* scores) of the remaining interaction terms for the first RDA axis then represent the extent of the difference between the particular treatment and the control, at a particular time point. The canonical coefficients must be transformed, however, to undo the standardization which CANOCO performs on the environmental variables and the species data. The actual formula for transforming the *Regr* scores can be found in the CANOCO reference manual (Ter Braak & Šmilauer 2002, section 8.3.10), but the necessary calculations can be performed by CanoDraw for Windows (using the *Project > Import variables > Setup PRC scores* command). There are as many response curves originating from the first RDA axis as there are treatment levels, in addition to control treatment.

Because the effect of experimental treatments at different sampling times cannot usually be summarized with only one constrained axis, we can construct second or even a further set of principal response curves. Note that the effect represented by the first-order and the additional response curves should be ascertained using a permutation test on the first canonical axis (or using the method outlined in this chapter, Section 9.1, to test for second and further constrained RDA axes). In the permutation test, we should randomly reallocate the experimental treatment of each whole time series. That means that all the samples taken at a particular permanent site 'travel' together during each permutation. The easiest way to assure this is to specify in the CANOCO program the individual time series (all the samples, taken through time at one site) as whole plots and ask for random permutations of the whole plots and no permutation of the split-plots.

The construction and testing of PRC is illustrated in Case study 5, in Chapter 15.

9.4.   Linear discriminant analysis

In some situations, we have an *a priori* classification of the studied objects (individuals of various species, sampling plots for vegetation description, etc.) and we want to find a quantitative classification rule using the values of measured (explanatory) variables to predict membership of an object in one of the pre-specified classes. This is a task for classical discriminant analysis. We will focus here on discriminant analysis seen as an ordination method that best reflects group membership.

Fisher's linear discriminant analysis (LDA), also called canonical variate analysis (CVA, used for example in CANOCO reference manual, Ter Braak & Šmilauer 2002), is a method that allows us to find the scores for the classified objects (i.e. the samples in CANOCO terminology). The object scores are expressed as linear combinations of the explanatory variables that optimally separate the *a priori* defined groups. The method is available in Canoco for Windows with some additional features not available in the standard implementations of this method.

To perform LDA in CANOCO, the classification of samples must be used as the species data. Each variable then represents a single class and the samples belonging to a class have a value of 1.0 for this variable and zero values for the other variables. This is the appropriate coding for classical discriminant analysis, where the classification is 'crisp'.*

The variables we want to use for the discrimination enter the analysis in CANOCO as environmental variables. We then select a canonical correspondence analysis (CCA) using Hill's scaling with a focus on the species distances (value −2 in the console version of CANOCO).

---

* CANOCO allows us to perform a discriminant analysis based on a 'fuzzy' classification, where (some) samples can belong to several classes at the same time. This situation might represent a true partial membership in several classes or our uncertainty about the true membership for those samples. The only requirement for fuzzy coding is that the sum of sample values in the species data is equal to one.

The one distinct advantage of doing LDA in CANOCO is that we might perform a **partial** discriminant analysis. In such an analysis, we can look for explanatory variables allowing us to discriminate between given classes in addition to other known discriminatory variables.

Other distinct features of LDA in CANOCO are the ability to select a subset of the discriminating (explanatory) variables by means of forward selection of environmental variables and the ability to test the discriminatory power of the variables by the non-parametric Monte Carlo permutation test.

When plotting the results, the species scores represent the means (centroids) of the individual classes in the discriminatory space. Scores of samples (the *SamE* scores) are the discriminant scores for the individual observations. A biplot diagram containing biplot scores of environmental variables (*BipE*) presents the table of averages of individual explanatory variables within individual classes, while the regression/canonical coefficients (*Regr* scores) of environmental variables represent the loadings of the individual variables on the discriminant scores.

Note that the lengths of the arrows for discriminating descriptors (plotted as the biplot scores of environmental variables) do not correspond properly to the discriminatory power of the particular variable. The CANOCO manual suggests in Section 8.4.3.1 how to rescale the length of individual arrows to reflect their efficiency for separating the *a priori* classes. The necessary steps are also illustrated in Case study 7, Chapter 17.

10

Visualizing multivariate data

The primary device for presenting the results of an ordination model is the ordination diagram. The contents of an ordination diagram can be used to approximate the species data table, the matrix of distances between individual samples, or the matrix of correlations or dissimilarities between individual species. In ordination analysis including environmental variables, we can use the ordination diagram to approximate the contents of the environmental data table, the relationship between the species and the environmental variables, the correlations among environmental variables, etc. The following two sections summarize what we can deduce from ordination diagrams that result from linear and unimodal ordination methods.

Before we discuss rules for interpreting ordination diagrams, we must stress that the absolute values of coordinates of objects (samples, species, explanatory variables) in ordination space do not have, in general, any meaning.* When interpreting ordination diagrams, we use relative distances, relative directions, or relative ordering of projection points.

10.1. What we can infer from ordination diagrams: linear methods

An ordination diagram based on a linear ordination method (PCA or RDA) can display scores for samples (represented by symbols), species (represented by arrows), quantitative environmental variables (represented by arrows) and nominal dummy variables (represented by points – centroids – corresponding to the individual levels of a factor variable). Table 10-1 (after

---

* There are some exceptions to this statement. In analyses with particular options selected, distances between sample points measure their dissimilarity in composition turnover (SD) units; in other cases one can project tips of arrows of environmental variables onto ordination axes to read the value of correlation between the variable and the sample scores on an ordination axis.

Table 10-1. *Relations between species, samples and environmental variables that can be read from an ordination diagram for a linear method for the two types of scaling of ordination scores*

Compared entities	Scaling 1 Focus on sample distances	Scaling 2 Focus on species correlations
Species × samples	(Fitted) abundance values in the species data	
Samples × samples	Euclidean distances among samples	n.a.
Species × species	n.a.	Linear correlations among species
Species × EV	Linear correlations between species and environmental variables	
Samples × EV	n.a.	Values of variables in the environmental data
EV × EV	Marginal effects of environmental variables on sample scores	Correlations among environmental variables
Species × nominal EV	Mean species abundances within sample classes	
Samples × nominal EV	Membership of samples in classes	
Nominal EV × nominal EV	Euclidean distances between sample classes	n.a.
EV × nominal EV	n.a.	Averages of environmental variables within the sample classes
Biplot with this scaling is called:	Distance biplot	Correlation biplot
Species arrow length shows:	Contribution of species to the ordination subspace definition	Approximate standard deviation of the species values
Length of the radius of 'equilibrium contribution' circle:	$\sqrt{d/p}$	$s_j\sqrt{d/p}$

EV means environmental variables. In the definition of the equilibrium contribution circle, d expresses the dimensionality of the displayed ordination spaces (usually the value is 2), p means the dimensionality of the whole ordination space (usually equal to the number of species), and s_j is the standard deviation of the values of jth species.

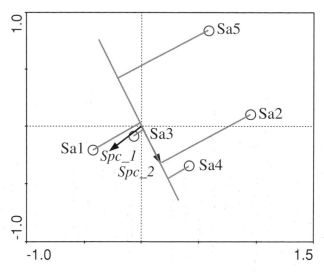

Figure 10-1. Projecting sample points onto species vector (of the *Spc_2* species) in a biplot from a linear ordination. Here we predict the largest abundance of *Spc_2* in samples Sa4 and Sa2, then sample Sa3 is predicted to have a lower abundance near the average of *Spc_2* values, and the expected abundance decreases even more for Sa1 and Sa5.

Ter Braak 1994) summarizes what can be deduced from ordination diagrams based on these scores. Additionally, the meaning of the length of species arrows in the two types of ordination scaling is discussed, and information about calculating the radius of the equilibrium contribution circle is given. In the ordination diagram the **equilibrium contribution circle** displays the expected positions of the heads of species arrows, under the assumption that the particular species contribute equally to the definition of all the ordination axes. Note that an equilibrium contribution circle has a common radius for all species only in a biplot diagram focusing on inter-sample distances (see Legendre & Legendre 1998, for additional discussion).

If we project the sample points perpendicular to a species' arrow, we obtain an approximate ordering of the values of this species across the projected samples (see Figure 10-1). If we use the sample scores that are a linear combination of the environmental variables[$] (*SamE* scores, typically in RDA), we approximate the **fitted**, not the observed, values of those abundances.[+] This interpretation is correct for both kinds of scaling. If centring by species was performed,[*]

[$] Program CanoDraw uses the *SamE* scores for samples in the ordination diagrams from direct gradient analyses (RDA, CCA). This default option can be changed, however.

[+] See Section 5.3 for a description of the linear ordination techniques in terms of fitting models of a species linear response.

[*] This is the case in the majority of analyses using linear ordination methods.

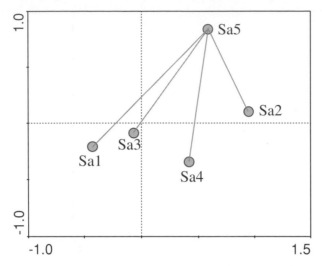

Figure 10-2. Distances between sample points in an ordination diagram. If we measured the dissimilarity between sample Sa5 and the other samples using Euclidean distance, the distance between sample Sa5 and Sa2 is predicted to be the shortest one, the distance to samples Sa3 and Sa4 the next shortest, and finally the distance to sample Sa1 the longest (i.e. samples Sa1 and Sa5 are predicted to show the greatest dissimilarity).

a sample point projecting onto the beginning of the coordinate system (perpendicular to a species arrow) is predicted to have an average value of the corresponding species. The samples projecting further from zero in the direction of the arrow are predicted to have above-average abundances, while the sample points projecting in the opposite direction are predicted to have below-average values.

The perpendicular projection of points onto the vectors (arrows) is called the **biplot rule** in this context.

Only in the scaling focused on sample distances does the distance between the sample points approximate their dissimilarity, expressed using Euclidean distance (see Figure 10-2).*

Only in the scaling focused on species correlations do the relative directions of the species arrows approximate the (linear) correlation coefficients among the species (Figure 10-3). In the case of covariance-based biplots (where species scores are not post-transformed) the approximated correlation between two variables is equal to the cosine of the angle between the corresponding

---

* Note that the actual distance measure approximated in an ordination diagram may often be different from the rough Euclidean distance. For example, if you specify standardization by sample norm in CANOCO, the distances between sample points approximate so called Chord distance (see Legendre & Gallagher 2001, and Section 6.2).

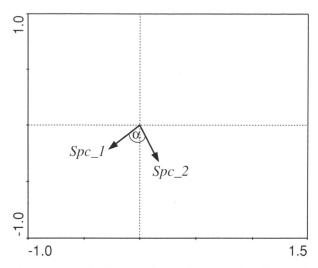

Figure 10-3. Angles between the species arrows in a diagram from a linear ordination method. As the arrows for the two species *Spc_1* and *Spc_2* meet nearly at right angles, the species are predicted to have a low (near-to-zero) correlation. More precise approximation in the default scaling options (with species scores being post-transformed) is achieved by the biplot projection rule.

arrows. Therefore, arrows pointing in the same direction correspond to species that are predicted to have a large positive correlation, whereas species with a large negative correlation are predicted to have arrows pointing in opposite directions.

If the species scores are post-transformed (divided by species standard deviation), we can estimate the correlations by perpendicularly projecting the arrow tips of the other species onto a particular species arrow. In most cases, we obtain very similar conclusions with both alternative interpretation rules.

We can apply a similar approximation when comparing the species and (quantitative) environmental variable arrows (see Figure 10-4). For example, if the arrow for an environmental variable points in a similar direction to a species arrow, the values for that species are predicted to be positively correlated with the values for that environmental variable. This interpretation is correct for both types of scaling of ordination scores.

The sample points can also be projected perpendicularly to the arrows of environmental variables (see Figure 10-5). This gives us the approximate order-ing of the samples in order of increasing value of that environmental variable (if we proceed towards the arrow tip and beyond it). The environmental variables (and covariables) are always centred (and standardized) before the ordination model is fitted. Thus, similar to projecting the sample points on the species

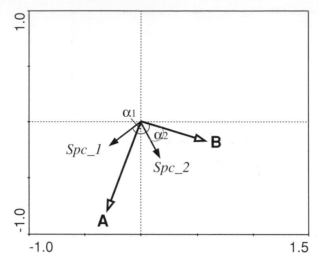

Figure 10-4. Angles between arrows of species and environmental variables in an ordination diagram from a linear method. When comparing the correlation of environmental variable B with the two species, we predict from the angles that B is slightly negatively correlated with species *Spc_1* and has a larger positive correlation with species *Spc_2* (the higher the value of B, the higher the expected value of *Spc_2*). The biplot projection of the species arrow tips onto the arrow of an environmental variable provides a more precise approximation.

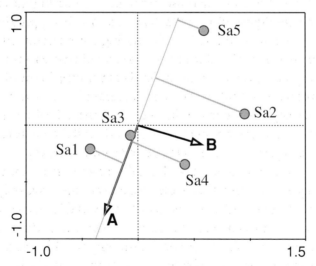

Figure 10-5. Projecting sample points onto the arrows of a quantitative environmental variable. Note that variable A is predicted to have similar values for the samples Sa3 and Sa4 even though they are located at different distances from the environmental variable line.

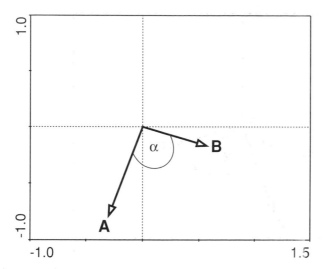

Figure 10-6. Measuring angles between arrows of quantitative environmental variables. The angle between the two variables suggests that they are almost non-correlated.

arrows, a projection point near zero (the coordinate system origin) corresponds to the average value of that particular environmental variable in that sample.[*]

The angle between the arrows of the environmental variables can be used to approximate the correlations among those variables in the scaling focused on species correlations (see Figure 10-6). Note, however, that this approximation is not as good as the one we would achieve if analysing the environmental data table as the primary data in a PCA. If the scaling is focused on inter-sample distances, we can interpret each arrow independently as pointing in the direction in which the sample scores would shift with an increase of that environmental variable's value. The length of the arrow allows us to compare the size of such an effect across the environmental variables (remember that all the environmental variables enter the analysis with a zero average and a unit variance).

CANOCO output allows a different (and usually a more useful) interpretation for the dummy environmental variables. These variables have only the 0 and 1 values and they are created by re-coding a factor variable. Such variables can be represented by symbols that are placed at the centroids of the scores for samples that have a value of 1 for the particular dummy variable. We can view the original factor variable as a classification variable and then the individual dummy variables correspond to the individual sample classes. We can say that

---

[*] Note, however, that the constrained ordination model is aimed at approximating the relations between species data and the environmental variables, rather than an optimal approximation of the environmental values for the individual samples.

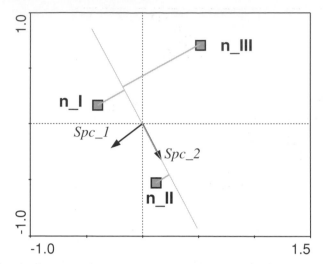

Figure 10-7. Projecting centroids of dummy environmental variables on species arrows in an ordination diagram from a linear method. Here we predict the largest average abundance of species *Spc_2* for the samples belonging to the class n_II, with samples from the n_I and n_III classes having a lower (and similar) average predicted abundance.

the centroid score for a dummy variable (referred to as *CenE* in the CANOCO documentation) represents the average of the scores of samples belonging to that class.

If we project the centroids of dummy environmental variables onto a species arrow, we can approximate the average values of this species in the individual classes (Figure 10-7). Similarly, the distance between the centroids of environmental variables approximates (in the scaling focused on inter-sample distances) the dissimilarity of their species composition, expressed using Euclidean distance (i.e. how different is the species composition of the classes).*

In both types of scaling, the distance between the centroids of individual sample classes and a particular sample point allows us to predict membership of that sample (Figure 10-8). A sample has the highest probability of belonging to the class with its centroid closest to that sample point. We must note that in a constrained analysis, where the dummy variables are used as the only explanatory variables, the constrained sample scores (*SamE* scores) have identical coordinates to the *CenE* scores of the classes to which the samples belong.

---

* Note, however, that if we do compare the positions of centroids in a constrained ordination model (RDA or, with the unimodal model, CCA), the underlying ordination model is 'optimized' to show the differences between the classes. An unbiased portrait can be obtained from an unconstrained ordination with the centroids of dummy environmental variables *post hoc* projected into the ordination space.

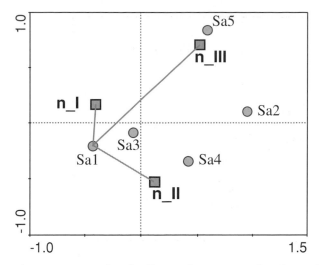

Figure 10-8. Measuring the distance between sample points and centroids of dummy environmental variables. We can predict here that sample Sa1 has the highest probability of belonging to class n_I, and has the lowest probability of belonging to class n_III.

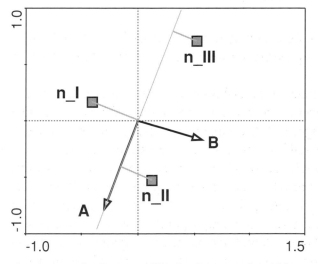

Figure 10-9. Projecting centroids of environmental variables onto arrows of quantitative environmental variables. The samples from class n_II are predicted to have the largest average value of variable A, followed by samples from class n_I; the samples from n_III have the lowest average value of the variable A.

If we project the centroids of dummy environmental variables onto an arrow of a quantitative environmental variable, we can deduce the approximate ordering of that variable's average values in the individual sample classes (see Figure 10-9).

Table 10-2. *Relation between species, samples and environmental variables that can be read from an ordination diagram of the weighted-averaging (unimodal) ordination method for two types of scaling of ordination scores*

Compared entities	Scaling 1 Focus on sample distances and Hill's scaling	Scaling 2 Focus on species distances and biplot scaling
Species × samples	(Fitted) relative abundances of the species data table	
Samples × samples	Turnover distances among samples	χ^2 distances among samples (if eigenvalues comparable)
Species × species	n.a.	χ^2 distances among species distributions
Species × EV	Weighted averages – the species optima with respect to particular environmental variable	
Samples × EV	n.a.	Values of environmental variables in the samples
EV × EV	Marginal effects of environmental variables	Correlations among environmental variables
Species × nominal EV	Relative total abundances in sample classes	
Samples × nominal EV	Membership of samples in the classes	
Nominal EV × nominal EV	Turnover distances between sample classes	χ^2 distances (if λs comparable) between sample classes
EV × nominal EV	n.a.	Averages of environmental variables within sample classes
Species-samples diagram with this scaling is called:	Joint plot	Biplot

EV means environmental variables. Lambda represents the eigenvalue of the considered ordination axis.

10.2.   What we can infer from ordination diagrams: unimodal methods

The interpretation of ordination diagrams based on a unimodal ordination model is summarized in Table 10-2 (following Ter Braak & Verdonschot 1995). It has many similarities with the interpretation we discussed in detail for the linear ordination model in the preceding section, so we will point the reader to the preceding section when required.

The main difference in interpreting ordination diagrams from linear and unimodal ordination methods lies in the different model of the species response along the constructed gradients (ordination axes). While a linear (monotonic) change was assumed in the preceding section, here (many of) the species are assumed to have an optimum position along each of the ordination axes with their abundances (or probability of occurrence for presence–absence data) decreasing symmetrically in all directions from that point.* The estimated position of that species' optimum is displayed as its score, i.e. as a point (symbol). These positions are calculated as the weighted averages of the sample positions with weights related to the species' abundances in the respective samples.

Another important difference is that the dissimilarity between samples is based on the chi-square metric, implying that any two samples with identical **relative** abundances (say two samples with three species present and their values being 1 2 1 and 10 20 10, respectively) are judged to be identical by a unimodal ordination method. The dissimilarity of the distribution of different species is judged using the same kind of metric, being applied to a transposed data matrix.

The mutual position of sample and species points allows us to approximate the relative abundances in the species data table. The species scores are near the points for samples in which they occur with the highest relative abundance and, similarly, the sample points are scattered near the positions of species that tend to occur in those samples (see Figure 10-10). This kind of interpretation is called the **centroid principle.** Its more quantitative form works directly with

---

* The question about how many species in one's data should show a linear or a unimodal response with respect to the environmental variables frequently arises among CANOCO users in trying to decide between linear and unimodal ordination methods. The key fact for answering this is that the ordination model is simply a **model.** We must pay for the generalization it provides us with some (sometimes quite crude) simplification of the true community patterns. We should try to select the type of ordination model (linear or unimodal), improved by an appropriate transformation of the species data, which fits better than the alternative one, not the one which fits 'perfectly'. While this might sound like a highly disputable view, we do not think that data sets where the choice of one or the other ordination type would be inappropriate occur very often.

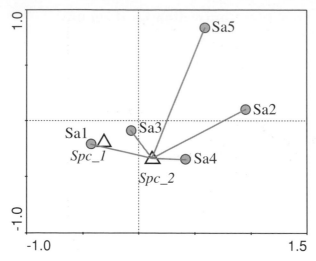

Figure 10-10. The distance between a species point and the sample points. The species *Spc_2* is predicted to have the highest relative abundance in the samples Sa3 and Sa4 and the lowest one in sample Sa5 (it is probably absent from this sample).

distances between points. If we order the samples based on their distance to a point for a particular species, this ordering approximates the ordering based on the decreasing relative abundance of that species in the respective samples.

For shorter gradient lengths (less than 2 SD units, approximately) we can interpret the positions of species and samples in the ordination plot using the **biplot rule** (see Figure 10-11). This is similar to the interpretation used in the ordination diagrams based on the linear ordination methods. We simply connect the species points to the origin of the coordinate system and project the sample points perpendicular to this line.[*]

The distance between the sample points approximates the chi-square distances between samples in the biplot scaling with the focus on species, but only if the ordination axes used in the ordination diagram explain a similar amount of variability (if they have comparable eigenvalues).

If we use Hill's scaling with its focus on inter-sample distances, then the distance between the samples is scaled in 'species turnover' units[X] that correspond to the labelling of the ordination axes. The samples that are at least four units apart have a very low probability of sharing any species, because a 'half change' distance in the species composition is predicted to occur along one SD unit.

The distance between species points in the biplot scaling (with its focus on the species distances) approximates the chi-square distance between the species distributions (see Figure 10-12).

---

[*] In DCA, use the centroid of sample scores instead of the coordinates' origin.

[X] These units are called **SD units**, a term derived from the 'standard deviation' (i.e. width) of a species response curve (see Hill & Gauch 1980).

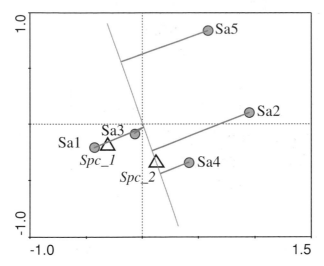

Figure 10-11. Biplot rule applied to species and sample points in an ordination diagram from a unimodal ordination method. The species *Spc_2* is predicted to have the highest relative frequency in samples Sa4 and Sa2 and the lowest one in sample Sa5. This species is predicted to occur in samples Sa3 and Sa1 with its average relative frequency.

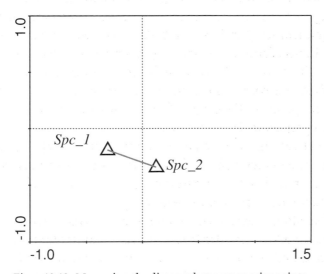

Figure 10-12. Measuring the distance between species points.

If we project species points onto an arrow of a quantitative environmental variable, we get an approximate ordering of those species' optima with respect to that environmental variable (see Figure 10-13). Similarly, we can project sample points on the arrow for a quantitative environmental variable to approximate values in the environmental data table, but only if biplot scaling is used.

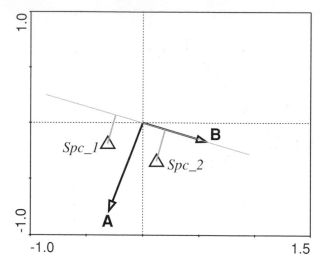

Figure 10-13. Projecting species points onto arrows of quantitative environmental variables. We can interpret the graph by saying that species *Spc_2* is predicted to have its optimum with respect to environmental variable B at higher values of that variable than species *Spc_1*.

Interpretation of the relationships between environmental variable arrows (either using the angle between the arrows or comparing the relative directions and size of the impact) is similar to the interpretation used in linear methods, described in the preceding section (see also Figure 10-6).

For centroids of dummy environmental variables (see previous section for an explanation of their meaning), we can use the distance between the species points and those centroids to approximate the relative **total** abundances of the species in the samples of that class (summed over all the samples in the particular class) – see Figure 10-14.

Comparison between the sample points and centroids of dummy variables and between the centroids and the arrows for the quantitative environmental variables proceeds as in linear models, as described in the previous section.

The distance between the particular centroids of dummy environmental variables is interpreted similarly to the distance (dissimilarity) between the sample points. In this case the distance refers to the average chi-square distances (or turnover, depending on the scaling) between the samples of the corresponding two sample classes.

10.3.   Visualizing ordination results with statistical models

The results of ordination methods can be summarized in one or a few ordination diagrams; in addition, the determined compositional gradients (ordination axes) can be used as a framework within which we can study

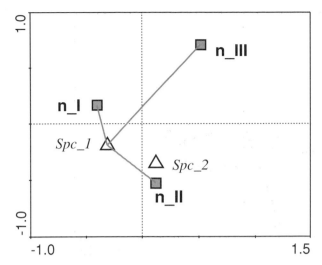

Figure 10-14. Measuring the distance between species points and centroids of dummy environmental variables. The average relative frequency of species *Spc.1* is predicted to be highest in class n_I, followed by n_II and then by class n_III.

the behaviour of individual variables, various sample characteristics (such as species richness or diversity), or even the relationship between various variables.

The relationships between various variables (including independent variables, as well as the scores of samples or species on the ordination axes) can be abstracted ('formalized') with the help of regression models and their use is illustrated in the tutorial in Section 8.7. Program CanoDraw also contains other visualization methods, which we illustrated here.

You can create, for example, a plot displaying the values of a particular environmental variable in ordination space. The actual value of the selected variable in a sample is displayed at the sample position using a symbol with its size proportional to the value. If we expect the value of the environmental variable to change monotonically across the ordination (sub)space (or even linearly, as in the constrained ordination model), we might recognize such a pattern from this **symbol plot** (see Figure 10-15). However, plotting individual sample values for even a medium-sized data set often does not allow for an efficient summary of the pattern. In such case, some kind of regression model should be used (GLM, GAM, or loess smoother, see Chapter 8).

10.4.   Ordination diagnostics

Ordination diagrams and attribute plots can also be used to check how well the data analysed with an ordination method fulfil the assumptions of the underlying model.

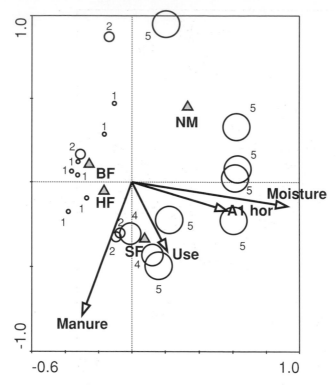

Figure 10-15. Attribute plot with symbols, illustrating the variation in the values of the Moisture variable. BF, NM, SF and HF are nominal (qualitative) environmental variables, while Use, A1 hor, Moisture and Manure are (semi-)quantitative environmental variables.

The first obvious check concerns our assumption of the shape of the species response along the ordination axes, which represent the 'recovered' gradients of the community composition change.* We can fit regression models describing the change of species values along the ordination axes with the *Create > Attribute Plots > Species response curves* command. In the case of both linear and unimodal ordination methods, we should probably let CanoDraw select the regression model complexity, whether we use generalized linear or generalized additive models. The steps needed, as well as additional considerations, are given in Section 8.7.

Another assumption of constrained ordination methods (both RDA and CCA) is that the gradients of community composition change depend on the environmental descriptors in a linear way. This is enforced by the constrained ordination methods, where the constrained ordination axes are defined as

---

* With an underlying environmental interpretation in the case of constrained ordination methods.

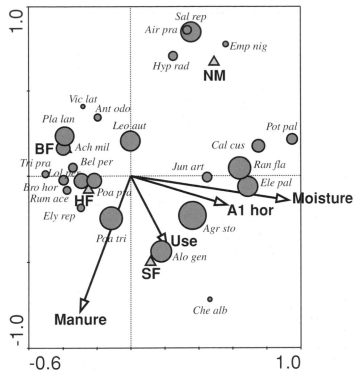

Figure 10-16. Attribute plot showing the values of the fit of species into the ordination space (the *CFit* statistics in CANOCO output).

linear combinations of the submitted environmental variables. Nevertheless, often a simple monotone transformation of a predictor can lead to its substantially higher predictive power. This happens, for example, if you study the successional development of (plant or other kind of) a community during secondary succession. The pace of community change is usually highest at the beginning and slows down continually. If you use the time since the start of succession as a predictor, the community change is usually better explained by the log-transformed predictor. This corresponds to a non-uniform response of community composition along the successional time gradient.

The CANOCO manual refers to **ordination diagnostics** with a somewhat different meaning; namely, it refers to the statistics calculated with the ordination method. These statistics describe, in general terms, how well the properties of individual samples and of individual species are characterized by the fitted ordination model. This statistic is named *CFit* in the case of species (see Ter Braak and Šmilauer 2002, Section 6.3.11.2) and can be used in CanoDraw to select which species occur in the ordination diagram. Additionally, we can plot these values in the ordination space (see Figure 10-16).

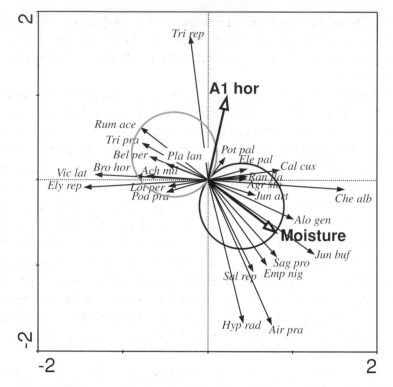

Figure 10-17. An example of t-value biplot. *Jun art* is *Juncus articulatus*; *Ran fla* is *Ranunculus flammula*, *Pla lan* is *Plantago lanceolata*, *Ach mil* is *Achillea millefolium* and *Lol per* is *Lolium perenne*.

10.5.   t-value biplot interpretation

A t-value biplot is a diagram containing arrows for the species and arrows or symbols for the environmental variables. It is primarily used to reveal statistically significant pair-wise relationships between species and environmental variables (i.e. which species depend on which environmental variables). We interpret the t-value biplot using the **biplot rule** (see Section 10.1). Using the biplot projections, we approximate the t-values of the regression coefficients, which we would get for a multiple regression with the particular species acting as a response variable and all the environmental variables as predictors.

To interpret the t-value biplot (see Figure 10-17 for an example), we project the tips of the environmental variable arrows onto a line overlying the arrow of a particular species. If the environmental variable arrow's tip projects on the line further from the origin than the species' arrowhead, we deduce that the t-value of the corresponding regression coefficient is larger than 2.0. This also holds true for projection onto the part of the line going in the opposite direction, with the t-value being less than −2.0. Projection points between these limits (including the coordinate system origin) correspond to t-values between −2 and +2, with the predicted value of a T statistic being 0.0 for the coordinates' origin.

If we want to find which species have a significant **positive** relationship with a particular environmental variable, we can draw a circle with its centre mid-way between the arrowhead for that variable and the origin of the coordinate system. The circle's diameter is equal to the length of that variable's arrow. Species lines that end in that circle have a positive regression coefficient for that environmental variable with the corresponding t-value larger than 2.0. We can draw a similar circle in the opposite direction, corresponding to a significant negative regression coefficient. These two circles are so-called **Van Dobben circles** (see Ter Braak & Šmilauer 2002, Section 6.3.12) and they are illustrated in Figure 10-17 for the variable *Moisture*. We can deduce (under the assumption of a sufficient number of samples – data points) that increasing moisture positively affects species such *Jun art* or *Ran fla*, and negatively affects the abundance of species such as *Pla lan*, *Ach mil* or *Lol per*. Note that there are as many independent pairs of Van Dobben circles as there are explanatory (environmental) variables.

Additional information on t-value biplots can be found in Ter Braak & Šmilauer (2002), Section 6.3.12.

Case study 1: Variation in forest bird assemblages

The primary goal of the analyses demonstrated in this case study is to describe the variability of bird communities, related to the differences in their habitat.

The data set originates from a field study by Mirek E. Šálek et al. (unpublished data) in Velka Fatra Mts. (Slovak Republic) where the bird assemblages were studied using a grid of equidistant points placed over the selected area of montane forest, representing a mix of spruce-dominated stands and beech-dominated stands. There was a varying cover of deforested area (primarily pastures) and the individual quadrats differed in their altitude, slope, forest density, cover and nature of shrub layer, etc. (see the next section for description of environmental variables). The primary data (species data) consist of the number of nesting pairs of individual bird species, estimated by listening to singing males at the centre points of each quadrat. The data values represent an average of four observations (performed twice in each of two consecutive seasons).

11.1. Data manipulation

The data are contained in the Excel spreadsheet file named *birds.xls*. This file has two sheets. We will use the data contained in the first sheet (named *birds*), where the bird species observed in only one of the quadrats were omitted (the full data are available in the other sheet, labelled *birds-full*). Note that the *birds* sheet contains both the species data (average abundance of individual bird species) and the environmental data (description of habitat characteristics for the particular quadrat). The species data are contained in the columns *A* to *AL* (column *A* contains sample labels and then there are data for 37 species) and the environmental data in the columns *AN* to *BA* (column *AN* contains sample labels again, followed by 13 environmental variables).

There is only one quantitative variable, *Altit*, specifying the average altitude of the sampled quadrat in metres above sea level. The next eight columns represent a mixture of semi-quantitative and strictly ordinal variables, describing: forest cover in the whole quadrat (*Forest*), average density of forest stands (*ForDens*), relative frequency of broad-leaved tree species in the tree layer (*BrLeaf*, with 0 value meaning spruce forest, and 4 only broad-leaved trees), total shrub layer cover (*E2*), percentage of coniferous species (spruce) in the shrub layer (*E2Con*), cover of the herb layer (*E1*) and its average height (*E1Height*). The variable *Slope* has a more quantitative character; it corresponds to slope inclination in degrees, divided by five. The last four variables represent two pairs of dummy variables coding two possible levels of two factors: the presence of larger rocks in the quadrat (*Rocks* and *NoRocks*) and the position of the quadrat on the sunny (south-east, south, and south-west) slopes (*Warm* and *Cold*).

To analyse these data with CANOCO, you must transform them into text files with CANOCO format. The species and environmental data must be processed separately and stored in independent data files. First select the rectangular area in the *birds* sheet representing the species data (columns *A* to *AL* and rows *1* to *44*). Copy the selected region to the Windows Clipboard (e.g. using the *Edit > Copy* command) and store the data in a file with the CANOCO format named, for example, *bird_spe.dta*, using the WCanoImp program, as described in Section 1.7. You can use either the condensed or the full format. Similarly, the environmental data stored in columns *AN* to *BA* and rows *1* to *44* should be exported into the file named *bird_env.data*. Parts of the contents of the two files are illustrated in Table 11-1.

11.2.   Deciding between linear and unimodal ordination

Start your analysis with the decision about the appropriate kind of ordination model, using the method outlined in Section 4.3. Note again that this procedure is recommended only when you do not have an indication of which of the two families of ordination methods to use, based either on the peculiarities of your data set or on your experience. In our particular example, it is difficult to decide *a priori* about the method and so it is good to estimate the heterogeneity in the species data, using the length of the community composition gradients in species turnover units, as calculated by detrended correspondence analysis. Define a new CANOCO project and in the project setup wizard make the following choices:

1. You have *Species and environmental data available*. While you do not need the environmental variables in this step, you can acknowledge their

Table 11-1. *Part of the contents of the two data files used in this case study*

Part of the created species data file

```
Bird species assemblages in montane forest, Velka Fatra, Slovak Rep.
(I5,1X,6(I6,F5.1))
6
      1    9  1.0   12  2.0   14  1.0   17  1.0   19  1.0   22  1.0
      1   25  1.0   27  1.0
      2    9  1.5   12  3.0   13  0.5   14  2.0   17  0.8   19  0.5
      2   21  0.5   22  1.0   23  0.5   24  1.0   27  0.5   29  0.8
      2   32  0.3
      3    7  1.0    9  1.0   12  3.0   14  2.0   15  0.5   22  0.5
      3   23  0.5   24  1.0   26  0.5   27  0.5   30  0.5   32  0.5
      3   33  0.3   34  0.5   35  0.5
      4    9  1.5   12  2.8   14  1.0   22  0.5   28  0.5   29  0.5
      4   32  0.5   34  0.3   35  0.5
...
     43    3  1.0    9  2.0   12  1.5   14  1.0   19  1.5   20  0.5
     43   21  1.0   22  1.0   24  0.5   27  1.0   28  0.5   29  0.5
      0

AegiCaudAlauArveAnthTrivCertFamiColuPaluCorvCoraCucuCanoDryoMartEritRubeFiceAlbi
FiceParvFrinCoelLoxiCurvParuAterParuCrisParuMajoParuMontPhoePhoePhylCollPhylSibi
PhylTro PrunModuPyrrPyrrReguReguReguIgniSittEuroSylvAtriTrogTrogTurdMeruTurdPhil
TurdViscTurdTorqCardSpinRegu_sp.ParuPaluMotaCineOriOri
R1        R2        R3        R4        R5        R6        R7        R8        R9        R10
R11       R12       R13       R14       R15       R16       R17       R18       R19       R20
R21       R27       R28       R29       R30       R31       R32       R37       R38       R39
R40       R41       R42       R43       R46       R47       R48       R49       R50       R51
R54       R55       R56
```

Part of the created environmental data file:

```
Birds data - habitat characteristics, Velka Fatra Mts., Slovak Rep.
(I5,1X,13F5.0)
13
      1    860   4   3   2   2   2   3   1   8   0   1   0   1
      2   1010   4   3   2   0   0   1   1   5   1   0   0   1
      3   1070   4   2   2   2   2   4   1   8   0   1   0   1
      4   1050   4   2   2   2   2   3   1   9   0   1   0   1
      5   1030   4   2   2   2   2   6   1   7   0   1   0   1
...
     42   1050   4   2   3   0   0   9   2   7   0   1   1   0
     43   1045   1   2   4   3   1   9   2   8   0   1   1   0
      0      0   0   0   0   0   0   0   0   0   0   0   0   0
```

Altit	Forest	ForDens	BrLeaf	E2	E2Con	E1	E1Height	Slope	Rocks
NoRocks	Warm	Cold							
R1	R2	R3	R4	R5	R6	R7	R8	R9	R10
R11	R12	R13	R14	R15	R16	R17	R18	R19	R20
R21	R27	R28	R29	R30	R31	R32	R37	R38	R39
R40	R41	R42	R43	R46	R47	R48	R49	R50	R51
R54	R55	R56							

availability here, because you will use this project setup as the starting point of your other CANOCO projects in this case study. At the bottom of the first setup wizard page, select the *indirect gradient analysis* option, however.

2. On the next page, locate the source data files with species and environmental data and also specify a name for the file containing CANOCO results and the directory where it will be placed. Note that for a new analysis CANOCO offers the software's installation directory after the *Browse* button is pressed, which might not be appropriate, particularly if you work with a networked installation.

3. On the next setup wizard page, select the DCA method and *by segments* in the *Detrending Method* page. On the next page, use *Log transformation*.*

4. Do not check any option on the following page (*Data Editing Choices*). Pressing the *Next* button here brings you to the last setup wizard page and you conclude the setup by selecting the *Finish* button there. The resulting analysis must be saved now (e.g. under the name *dca.con*) and you can run the analysis using the *Analyze . . .* button in Project View.

5. You should now switch to the Log View to see the results. The first two axes explain approximately 23% of the variability in the species data. More important at this stage is the line giving the lengths of recovered composition gradients (DCA axes):

```
Lengths of gradient : 2.001 1.634 1.214 1.613
```

The gradients are not very long, so you are advised to use the linear ordination methods for analyses.

11.3. Indirect analysis: portraying variation in bird community

1. Start with an indirect analysis of the species data, using PCA as the method to summarize the community variation. Even when the unconstrained analysis does not need the environmental variables, we keep them in the analysis – in an indirect analysis they are passively projected into the resulting ordination space and can suggest interpretations for the principal components. You can start a new CANOCO project using the previous one, where several choices (names of the input data files, transformation of species data) remain appropriate for this analysis. With the *dca.con* project opened in the Canoco for

---

* The alternative square-root transformation would work as well, with the additional bonus of escaping the problem of the constant we must add to the data values before calculating the logarithm, to prevent taking the log of the zero values.

Windows workspace, select the *File > Save As . . .* command from the menu and specify a new name for your CANOCO project (*pca.con* is suggested here). Select *Clear the log window* when prompted by the CANOCO program.

2. Now click the *Options . . .* button in the Project View to invoke the setup wizard again. On the first page, leave the *Species and environment data available* choice, as well as the *interpret patterns extracted from all variation (indirect gradient analysis)*. On the next page, you should change the name of the CANOCO results file (*pca.sol* is suggested here), and then change from *DCA* to *PCA* on the next page. The options on the *Scaling:Linear Methods* wizard page have default values appropriate for your analysis.* Keep the transformation options as they were defined in the test project with DCA (changing them would partly invalidate the conclusions made in the previous section).

3. The choices to be made in the next setup wizard page (*Centering and Standardization*) are crucial for the interpretation of the PCA results. In the right half of this page, the default value is *Center by species*. The centring is appropriate in almost any linear analysis, while the standardization (most often to be performed in addition to centring) is not so easy. In this case, if you stress the fact that the abundances of all species are in the same units (counts of nesting pairs over comparable areas) and are therefore comparable, you might prefer not to standardize. If, on the other hand, you take the view that different bird species with different territory size, sociability and feeding habits differ in their potential frequency of occurrence in the landscape, you might prefer to make the counts of different species comparable by their standardization. The choice is not easy, but in this tutorial follow the 'no standardization, centre-only' path. Further consideration for this choice is that the log transformation makes the frequencies comparable, to some extent, as it lets the analysis focus on relative change (see also Section 1.8). For the choices on the left half of this wizard page, see more comments in Section 14.3, where linear ordinations with and without standardization by sample norms are compared. Your analysis aims to portray both the differences in the proportions taken by individual species and the difference in their absolute counts. So, leave the *None* option selected here (see Figure 11-1).

4. Leave the wizard page with *Data Editing Choices* at its default state (nothing selected there) and then close the project setup wizard sequence. Run the analysis in the same way as before and check the summary of the

---

* We are more interested in the correlation among the counts of individual species and between the species and environmental variables, rather than in the precise positioning of sample points in the ordination space.

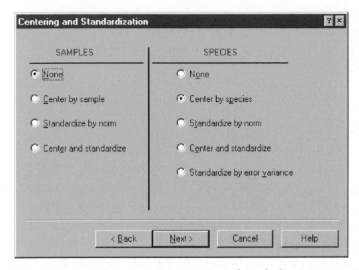

Figure 11-1. Resulting state of the *Centering and Standardization* setup wizard page.

analysis results in the Log View. The following text provides comments on selected parts of the CANOCO output displayed in this view. More information can be found in the Canoco for Windows manual (Ter Braak & Šmilauer 2002).

```
No samples omitted
 Number of samples          43
 Number of species          37
 Number of occurrences       533
```

This summary information (you have to scroll up in the Log View to find it) is a good place to check the correctness of your data sets and/or of the project choices. If the displayed numbers differ from those you expected, you should investigate the reasons for the discrepancies.

```
****** Check on influence in covariable/environment data ******
      The following sample(s) have extreme values
      Sample Environmental      Covariable   + Environment space
             variable Influence   influence      influence


         8          9       6.0x
        14          2       6.1x
        14          3       7.4x
      ****** End of check ******
```

The information in this section can be used for different purposes and one of them is checking for typing errors in the file with explanatory (environmental) variables. Because CANOCO uses regression to relate the variation in the species data to the values of environmental variables,[+] you can use the concepts of 'influence' or 'leverage' to identify outlying observations in your data sets. These observations correspond to samples with unusual values of environmental variables and/or covariables. In this example the output flagged sample 8 as unusual for the environmental variable 9 (i.e. the variable *Slope*). If you check the file with environmental data, you can see that in sample 8, the variable 9 has a value of 1, the smallest value found in the data set and the second smallest value is 3, not 2, so this quadrat on a flat part of the otherwise inclined study area differs quite a lot from the other ones.

Yet, after comparing the value with the lab protocols, it was concluded that this is a correct value. In many other real-world situations such an indication can, in fact, correspond to a typing error in the spreadsheet, where the decimal point was incorrectly shifted right or left from its true position (such as entering value 21 instead of 2.1). Therefore, it is worthwhile to check all the indicated cases before proceeding with the analysis. The other two 'spotted' outlying values are for the two variables describing forest cover and forest density in quadrat 14. Again, this quadrat is really different from the other ones, as it is the only one without any forest cover (i.e. only with the sub-alpine pastures).

Most interesting is the Summary table at the end of the CANOCO output.

Axes		1	2	3	4	Total variance
Eigenvalues	:	0.207	0.132	0.094	0.074	1.000
Species-environment correlations	:	0.886	0.778	0.675	0.603	
Cumulative percentage variance						
of species data	:	20.7	33.9	43.3	50.8	
of species-environment relation	:	36.8	54.8	64.5	70.6	
Sum of all eigenvalues						1.000
Sum of all canonical eigenvalues						0.442

You can see from this that the first two PCA axes (principal components) explain 33.9% (0.207 + 0.132) of the variability in species data. You can see that the amount of variability explained by individual axes decreases gradually, so you must face the difficult problem of how many ordination

---

[+] As part of the ordination space calculation in the case of direct gradient analysis and as a *post hoc* interpretation in the case of indirect gradient analysis, see Sections 3.5 and 3.6 for further details.

axes to present and interpret. One approach is to select all the ordination axes explaining more than the average variability explained per axis. This threshold value can be obtained by dividing the total variability (which is set to 1.0 in linear ordination methods in CANOCO) by the total number of axes. In our linear unconstrained analysis, the total number of axes is equal to the smaller value of the number of species and of the number of samples decreased by one, i.e. min $(37, 43 - 1) = 37$. This means that using this approach we would need to display and interpret all the principal components with eigenvalues larger than 0.0270, therefore not only the first four principal components, but possibly more. The results from this approach probably overestimate the number of interpretable ordination axes. Alternatively, the so-called broken-stick model is used (see Legendre & Legendre 1998, p. 409, for additional details). Using this approach, we compare the relative amount of the total variability explained by the individual axes, with the relative lengths of the same number of pieces into which a stick with a unit length would separate when selecting the breaking points randomly. The predicted relative length for the jth axis is equal to:

$$E(\text{length}_j) = \frac{1}{N} \sum_{x=j}^{N} \frac{1}{x}$$

where N is the total number of axes (pieces).

For the number of axes equal to 37, the broken-stick model predicts the values 0.1136, 0.0865, 0.0730 and 0.0640 for the first four principal components. Again, this implies that the fractions of variability explained by the four axes exceed values predicted by the null model and that all the axes describe non-random, interpretable variation in the species data. See Jackson (1993) for an additional discussion of selecting the number of interpretable axes in indirect ordination methods. For simplicity, we will limit our attention here to just the first two (most important) principal components.

The decision about the number of interpretable ordination axes is somewhat simpler for the constrained ordinations, where we can test the significance of all individual constrained (canonical) axes. Note that the distinction between the uncertainty in selecting an interpretable number of unconstrained axes and the much more clear-cut solution for the constrained axes stems from the semantic limitations of the word *interpretable*. In an unconstrained analysis, we do not provide a specific criterion to measure 'interpretability', while in the constrained one we look for interpretability **in relation to the values of selected explanatory variables**.

From the Summary table, you can also see that if you used the environmental variables to explain the variability in bird community composition, you

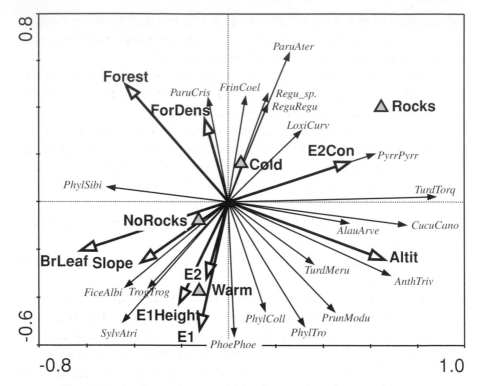

Figure 11-2. Species–environment biplot diagram from the PCA with environmental variables passively projected into the resulting ordination space.

would be able to explain up to 44.2% of the total variability. **Note, however, that you are performing an indirect analysis,** where the environmental variables are *post hoc* projected into the already determined ordination space.

The results of this analysis are best presented by a species–environmental variable biplot. The four dummy variables are displayed as centroids and the reader is advised to consult Chapter 10 for more information about visualizing the effects of categorical variables.

To create the biplot diagram shown in Figure 11-2, you must click the *CanoDraw* button and in the CanoDraw for Windows program (which was started by CANOCO) confirm saving a CanoDraw project file (*pca.cdw*) containing the information for creating diagrams for this project. To create the biplot, first specify the nominal variables *Rocks*, *NoRocks*, *Warm* and *Cold* (using the *Project* > *Nominal variables* > *Environmental variables* command), and then select *Create* > *Simple ordination plot* > *Species and environmental variables*. There are too many species arrows and labels in the resulting diagram, so you should display only the bird species, which are well characterized by the first two principal components. To do so, use the *Project* > *Settings* command and in the

Inclusion Rules page specify the value 15 in the *From* field of the *Species Fit Range* area, and close the dialog with the *OK* button. Then apply the new rule to your graph using the *Create > Recreate graph* menu command.

The first principal component is correlated mainly with altitude (increasing from the left to the right side of the diagram) and with the relative importance of broad-leaved trees in the tree layer (increasing in the opposite direction). Species such as *Turdus torquatus* or *Cuculus canorus* or *Anthus trivialis* tend to have larger abundance (or higher probability of occurrence) at higher altitude (their arrows point in similar directions as the arrows for *Altit*), and also in the spruce forest (their arrows point in opposite direction to the *BrLeaf* arrow). The second ordination axis is more correlated with the cover of the herb layer (*E1*), the height of the herb layer (*E1Height*), average forest density (*ForDens*), and the average cover of the shrub layer (*E2*). The *ForDens* versus the other descriptors have their values negatively correlated (because their arrows point in opposite directions).

The pairs of triangle symbols of the related dummy variables (*Rocks* and *NoRocks* or *Cold* and *Warm*) are not distributed symmetrically with respect to the coordinate origin. This is because their frequency in data is not balanced; for example, the samples that have rocks present (and therefore have value 1 for *Rocks* and 0 for *No Rocks*) are less frequent, so the corresponding symbol for *Rocks* lies further from the origin.

11.4. Direct gradient analysis: effect of altitude

There are several questions you can address with data sets concerning the relationship between explanatory variables and species composition. This tutorial is restricted only to detection of the extent of differences in composition of bird assemblages explainable by the quadrat average altitude (this section), and to studying the effects the other environmental variables have in addition to the altitudinal gradient (see the next section). To quantify the effect of altitude upon the bird community, you will perform a redundancy analysis (RDA), using *Altit* as the only environmental variable.

1. Start by saving the original CANOCO project with PCA under a new name: *rda1.con* is suggested.
2. Now you must make a few changes in the project setup. You invoke the project setup wizard, as in the previous section, by the *Options . . .* button. On the first wizard page, change the option at the bottom from *indirect gradient analysis* to *direct gradient analysis*. On the next page, change the name of the file containing the CANOCO results (*rda1.sol*). You can see on

the following wizard page (*Type of Analysis*) that the method was changed from *PCA* to *RDA* because of changes on the first page.

3. You do not need to make any changes on the following pages, up to the *Data Editing Choices* page. There, you should check the box for deletion of environmental variables. After you click the *Next* button, another page with the title *Delete Environmental Variables* appears. Here you select the variables 2 to 13 (all the variables except *Altit*) and transfer them to the right list using the >> button.

4. The next wizard page allows you to specify whether and in what way you want to select the retained environmental variables. Because we have only one environmental variable left, it does not make sense to subset it any further. Therefore, the *Do not use forward selection* option should remain selected.

5. The next page is the first one concerning Monte Carlo permutation tests and here you decide whether to perform the tests at all and what kind of test to use. You have only one explanatory variable, so there will be only one constrained (canonical) axis in this redundancy analysis. Therefore, CANOCO will produce the same results whether you choose the *Significance of first ordination axis* or the *Significance of all canonical axes* option. So select the test of the first ordination axis and in the right half of the page keep the value 499 as the number of permutations. On the next page, you select *Unrestricted permutations* and close the setup wizard on the following page using the *Finish* button.

6. After running the analysis, the resulting summary table appears in the Log View:

```
                                                              Total
Axes                                  1      2      3      4   variance
Eigenvalues                   : 0.115  0.152  0.118  0.075    1.000
Species-environment correlations: 0.792  0.000  0.000  0.000
Cumulative percentage variance
of species data               : 11.5   26.7   38.5   46.0
of species-environment relation : 100.0   0.0    0.0    0.0
Sum of all eigenvalues                                        1.000
Sum of all canonical eigenvalues                              0.115
```

The canonical axis (axis 1) explains 11.5% of the total variability in the species data and you can see that the first two of the following unconstrained axes explain individually more variability than the canonical axis (15.2% for axis 2, and 11.8% for axis 3). Nevertheless, the explanatory effect of the *Altitude* is significant.

This is confirmed by the report about the performed significance (Monte Carlo permutation) test, which follows the Summary table:

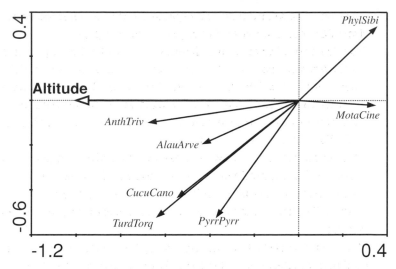

Figure 11-3. Species–environment biplot from RDA summarizing differences in bird assemblages along the altitudinal gradient.

```
**** Summary of Monte Carlo test ****

Test of significance of all canonical axes : Trace    =   0.115
                                             F-ratio  =   5.323
                                             P-value  =   0.0020

( 499 permutations under reduced model)
```

Note that the reported significance level estimate $(P = 0.0020)$ is the lowest achievable value given the number of permutations you used $[(0 + 1)/(499 + 1)]$. This means that none of the 499 ordinations based on the permuted data sets achieved as good a result as the 'true' one.

7. You can again summarize the results of this constrained analysis using the species–environment variable biplot (as shown in Figure 11-3). To do so, you must create a new CanoDraw project specific for this analysis (see previous section for summary of steps to be taken). Note that the diagram has 'mixed contents': the first (horizontal) axis is constrained, representing the variation in bird abundances explainable by the quadrat altitude, while the second (vertical) ordination axis is already unconstrained, representing the residual variation that is not explained by the *Altit* variable. It is, therefore, advisable to display in this biplot diagram only the species that have their abundances well explained by the first ordination axis, i.e. by the altitude.

Such species can be selected as species well fitted by the sample scores on the first ordination axis (see Section 5.3 for additional details about interpretation of PCA/RDA axes as regression predictors). Because the first RDA axis explains approximately 11% of the variability in species data, you can say that a species with an average 'explainability' by the first ordination axis will have at least 11% of the variability in its values explained by that axis. You can set this threshold in the CanoDraw program using the dialog box invoked by the *Project > Settings* menu command, using the *Inclusion Rules* page. Note that you must use the *Species* field in the *Lower Axis Minimum Fit*, because only the horizontal ('lower') axis is involved here. Note that only seven bird species pass this criterion.

11.5.   Direct gradient analysis: additional effect of other habitat characteristics

We will conclude this case study with a more advanced partial constrained ordination, which helps to address the following question:

> Can we detect any significant effects of the other measured habitat descriptors upon the bird community composition, when we have already removed the compositional variability explained by the average quadrat altitude?

This question can be addressed using a redundancy analysis where the *Altit* variable is used as a covariable, while the other descriptors are used as the environmental variables.

1. Start your analysis again by 'cloning' the previous one, saving the *rda1.con* project under the name *rda2.con*. Then, invoke the project setup wizard and start by changing the project settings on the first page. You should use covariables this time, so you must change the selection from the second to the third option in the upper part of this page. In the next page, you now have the third field enabled and you can copy there the contents of the second field, because you will use the same file for both the environmental variables and the covariables. You should also change the name of the solution file here (to *rda2.sol*).

2. Keep the choices on the following wizard pages up to the page *Data Editing Choices*. Here, you can see that setup wizard has automatically checked the deletion of covariables, because it is obvious that you need to differentiate the two sets of explanatory variables. After clicking the *Next* button, you must change the list of environmental variables to be deleted, actually swapping the contents of the two lists. You need to delete *only* variable 1 and *not* variables 2–13. On the next page (titled *Delete covariables*), you should delete variables 2–13 from the set of covariables

and keep only the *Altit* variable there. Then leave the *Do not use forward selection* option selected in the next page.

3. On the next wizard page, select *Significance of canonical axes together*, as you cannot *a priori* expect that one constrained axis is sufficient to describe the influence of the 12 environmental variables on the bird communities. On the next page, keep the option *Unrestricted permutations* selected, and do **not** check the option *Blocks defined by covariables* (our covariable is a quantitative descriptor and CANOCO would create a separate block for each unique altitude value). Finally, close the project setup wizard on the following page.

4. After the analysis, the resulting Summary table is displayed (with the other CANOCO output) in the Log View:

```
**** Summary ****

                                                              Total
Axes                                    1      2      3      4    variance

Eigenvalues                      : 0.112  0.067  0.042  0.034   1.000
Species-environment correlations : 0.886  0.810  0.818  0.765
Cumulative percentage variance
    of species data              : 12.6   20.3   25.0   28.8
    of species-environment relation: 34.2   54.7   67.4   77.8


Sum of all          eigenvalues                                0.885
Sum of all canonical  eigenvalues                              0.328
```

You can see that the amount of variability explained by the other habitat descriptors (in **addition to** the information provided by the quadrat altitude) is quite high (32.8% of the total variability in the bird counts data) compared with the variability explained by the altitude (11.5%, seen also from this Summary table, as the *Total variance − Sum of all eigenvalues* = 1.000 − 0.885). But you should realize that the current ordination model is much more complex (with DF = 10, given that variable 11 is linearly dependent on variable 10 and, similarly, variable 13 is linearly dependent on variable 12).

5. The Summary table for this analysis is followed by the report on the Monte Carlo permutation test. From it, you can see that the additional contribution of the 12 descriptors is highly statistically significant ($P = 0.0020$).

The species–environment biplot diagram summarizing the additional effects of the other habitat characteristics, when the altitude effects are already accounted for, is shown in Figure 11-4. Similar to the graph in the preceding section, only the 'well-fitting' species are included.

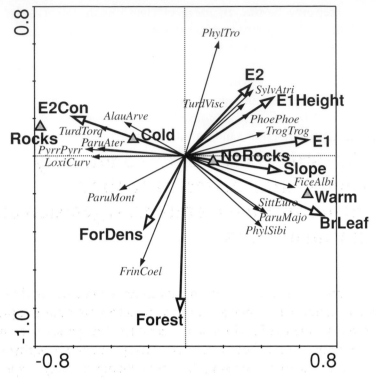

Figure 11-4. Species–environment biplot diagram summarizing the effects of habitat descriptors upon bird communities, after removing the altitudinal gradient.

We can see that the current first axis partly reflects the difference between the beech-dominated stands (on the right side) and the spruce forest (on the left side) and also the differences in the development of the herb layer (more developed in the sites on the right side of the ordination diagram). The second ordination axis correlates (negatively) with the total cover of the forest in the quadrat and, to a lesser extent, with average forest density.

12

Case study 2: Search for community composition patterns and their environmental correlates: vegetation of spring meadows

In this case study, we will demonstrate a common application of multivariate analysis: the search for a pattern in a set of vegetation samples with available measurements of environmental variables. The species data comprise classical vegetation relevés (records of all vascular plants and bryophytes, with estimates of their abundance using the Braun–Blanquet scale) in spring fen meadows in the westernmost Carpathian mountain ranges. The data represent an ordinal transformation of the Braun–Blanquet scale, i.e. the r, $+$, and 1 to 5 values of the Braun–Blanquet scale are replaced by values 1 to 7.[*] The relevés are complemented with environmental data – chemical analyses of the spring water, soil organic carbon content and slope of the locality. The ion concentrations represent their molarity: note that the environmental data are standardized in all CANOCO procedures and, consequently, the choice of units does not play any role (as long as the various units are linearly related).[+] The data are courtesy of Michal Hájek and their analysis is presented in Hájek et al. (2002). The aim of this case study is to describe basic vegetation patterns and their relationship with available environmental data, mainly to the chemical properties of the spring water.

The data are stored in the file *meadows.xls*, where one sheet represents the species data, and the other one the environmental data (in comparison with the original, the data are slightly simplified). The data were imported to CANOCO

---

[*] Note that the scale is not linear with respect to species cover and roughly corresponds to logarithmic transformation of cover, see Section 1.8.

[+] The environmental variables are not transformed, but the ion concentrations, which are always non-negative and often have a skew distribution, are log-transformed in such cases.

format, using the WCanoImp program. Two files were created, *meadows.spe* and *meadows.env*. Note that in the Excel file, the species data are transposed (i.e. species as rows, samples as columns). This will often be the case in similar broad-scale surveys, because the number of species is usually higher than the number of columns in ordinary spreadsheets. If both the number of species and the number of samples exceed the maximum number of columns, you can use either the CanoMerge program (see Section 4.1), or you must use some of the specialized programs for storing large databases, to create the data file with a CANOCO format (the data presented here were extracted from a database in the TURBOVEG program; Hennekens & Schaminee 2001). In fact, the analysed data set (70 samples, 285 species) is rather small in the context of recent phytosociological surveys.

12.1.   The unconstrained ordination

In the first step, you will calculate an unconstrained ordination – the detrended correspondence analysis (DCA). No data transformation is needed, because the ordinal transformation has a logarithmic nature with respect to cover and provides reasonable weighting of species dominance. In DCA with detrending by segments and Hill's scaling, the length of the longest axis provides an estimate of the beta diversity in the data set (the value 3.9 for our data set suggests that the use of unimodal ordination methods is quite appropriate here). The unconstrained ordination provides the basic overview of the compositional gradients in the data. It is also useful to include the environmental data in the analysis – they will not influence the species and samples ordination, but they will be projected afterwards to the ordination diagram.

Let's first inspect the summary in the log file:

```
Axes                                  1     2     3     4 Total inertia

Eigenvalues                      : 0.536 0.186 0.139 0.118        5.311
Lengths of gradient              : 3.945 2.275 2.049 2.015
Species-environment correlations : 0.898 0.409 0.531 0.632
Cumulative percentage variance
  of species data                :  10.1  13.6  16.2  18.4
  of species-environment relation:  26.4  28.4   0.0   0.0

Sum of all           eigenvalues                                 5.311
Sum of all canonical eigenvalues                                 1.560
```

You can see that the first gradient is by far the longest one, explaining about 10% of the total species variability (which is a lot, given the number of species in these data), whereas the second and higher axes explain much less. Also,

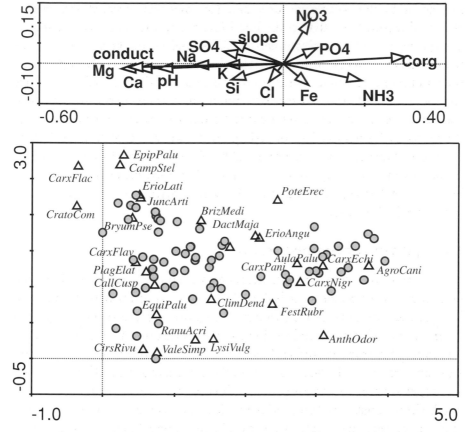

Figure 12-1. The species–samples (triangles and circles, respectively) biplot of the DCA of the whole data set in the lower diagram (only species with the highest weight are shown), and retrospective projection of the environmental variables in the upper diagram.

the first axis is very well correlated with the environmental data ($r = 0.898$), and the correlation for the other axes is considerably lower. All this suggests that the whole data set is governed by a single dominant gradient. The sum of all canonical eigenvalues in the printout corresponds to the sum of all canonical eigenvalues in the corresponding canonical analysis (i.e. it says how much could be explained by the environmental variables if they were used in similar, but constrained, analysis). The percentage variance of the species–environment relationship values represents percentages of this value. The number of axis scores calculated for a species–environmental variable biplot is restricted in a DCA, by default, to two. This is why the explained variability for the third and fourth axis is shown as 0.

Let us now create an ordination diagram. We suggest that you present the results in the form shown in Figure 12-1, i.e. a separate species–samples biplot, with an added plot of fitted environmental variables. It is not possible to include all the 285 species in the ordination diagram without total loss of clarity. In the DCA, the only possibility is to select species according to their weight, i.e. according to their sum over all the samples.* Consequently, only one-tenth of the total number of species is shown. Although this is probably the most feasible way to show the diagram in a publication, one should be aware that there is a substantial loss of information. You can accommodate more species in such a plot by not plotting the sample symbols (or plotting them in a separate scatterplot). Still, you will be able to plot only a minority of all the species. We recommend inspecting the CANOCO solution (*.sol*) file, to get information on more species. The other possibility is to create several scatterplots of species of varying frequency (in CanoDraw, the *Project* > *Settings* > *Inclusion rules* page enables you to first select the species with weight higher than 30%, then species with weight in the 25% to 30% interval, etc.). Journal editors will probably not allow you to include all the plots in your paper, but you can learn about the species behaviour from those plots. Another possibility for this data set is to put the bryophyte species into one plot, and the vascular plants into another one.

The samples are shown here without labels. Inspection of their general distribution suggests that there is a continuous variation of species composition in the whole data set and that we are not, therefore, able to find distinct vegetation types in the data set. The projection of environmental variables reveals that the first axis is negatively correlated with the *pH* gradient, with conductivity (*conduct*), and with the increasing concentration of cations (*Ca*, *Mg* and also *Na*), and positively correlated with soil organic carbon (*Corg*). The position of individual species supports this interpretation – with <u>Carex</u> <u>flacca</u>, or <u>Cratoneuron</u> <u>commutatum</u> being typical for the calcium-rich spring fens, and <u>Aulacomnium</u> <u>palustre</u>, <u>Carex</u> <u>echinata</u> and <u>Agrostis</u> <u>canina</u> for the acidic ones. The relationships of the species to the pH gradient are generally well known, and probably any field botanist with basic knowledge of local flora would identify the first axis with the pH gradient, even without any measured chemical characteristics.[†] The second axis is more difficult to interpret, as there are several variables weakly correlated with it.

The positions of arrows for environmental variables suggest that there is a group of variables that are mutually highly positively correlated (*pH*, *Ca*,

---

[*] Fit of species, which is another statistic useful for their selection into the ordination diagram, is not available for analyses with detrending by segments. You can, therefore, consider detrending by polynomials or a non-detrended analysis, and eventually define a group of well-fitting species there and import it into this DCA project.

[†] The situation might be rather different with other types of organisms, or in less known areas where such intimate knowledge of individual species ecology is not available.

*Mg, Na, conduct*ivity), and negatively correlated with organic carbon (*Corg*). However, you should recall that the diagram with environmental variables is based purely on their effect on species composition. A closer inspection of the correlation matrix in the CANOCO Log View shows that the variables are indeed correlated, but in some cases the correlation is not very great (particularly the negative correlation with organic carbon). The correlation matrix also confirms that the correlation of all the measured variables with the second axis is rather weak.

12.2.   Constrained ordinations

Now you can continue with the direct (constrained) ordinations. Whereas in DCA you first extract the axes of maximum variation in species composition, and only then fit the environmental variables, now you directly extract the variation that is explainable by the measured environmental variables. You will start with a canonical correspondence analysis (CCA) using all the available environmental variables. In the Global Permutation Test page of the setup wizard, specify that both types of permutation tests are performed. Both the test on the first axis and the test on all axes (on the trace) are highly significant ($P = 0.002$ with 499 permutations, which is the maximum under the given number of permutations). However, the F value is much higher for the test on the first axis ($F = 4.801$) than for the test on the trace ($F = 1.497$). This pattern also appears in the Summary table, where the first axis explains more than the second, third and fourth axes do together.

		1	2	3	4	Total inertia
Axes						
Eigenvalues	:	0.434	0.172	0.120	0.101	5.311
Species-environment correlations	:	0.921	0.831	0.881	0.880	
Cumulative percentage variance						
of species data	:	8.2	11.4	13.7	15.6	
of species-environment relation	:	27.8	38.9	46.6	53.1	
Sum of all eigenvalues						5.311
Sum of all canonical eigenvalues						1.560

You can compare this summary with that from the DCA you used before. You will notice that the percentage variance explained by the first axis is very close to that explained by the first axis in the unconstrained DCA (8.2 in comparison with 10.1), and also that the species–environment correlation is only slightly higher. This suggests that the measured environmental variables are those responsible for species composition variation. And indeed, in the ordination diagrams of DCA (Figure 12-1) and CCA (not shown here), the first axis of CCA is very similar (both for the species and for the sample scores) to the first axis of DCA. However, the second axes differ: the CCA shows a remarkable arch effect – the quadratic dependence of the second axis on the first one. Rather than trying the detrended form of CCA (i.e. DCCA), we suggest that the arch effect is caused by a redundancy in the set of explanatory variables (many highly correlated variables), and that you should try to reduce their number using a forward selection of environmental variables.

Before using the forward selection, it might be interesting to know whether the second axis is worth interpreting, i.e. whether it is statistically significant. To test its statistical significance, calculate the partial CCA with the environmental variables identical to those in the first CCA, and the sample scores (SamE, i.e. those calculated as linear combinations of the values of measured environmental variables) on the first axis used as the only covariable (see Section 9.1 for additional details). In this analysis, the variability explained by the original first axis is partialled out by the covariable, and the original second axis becomes the first one in this analysis, the third axis becomes the second one, etc. By testing the significance of the first axis in this analysis, you test, in fact, the second axis of the original analysis.

How do you prepare the file with the covariable? The easiest way is to open the solution file as a spreadsheet (e.g. in Excel) and copy the appropriate kind of sample scores to the Windows Clipboard. Be sure that you are using the SamE scores – they are at the end of the solution file. The copied scores must be in one column. Then, you will use the WCanoImp program to create a CANOCO-format data file.

It is worthwhile checking that the axes correspond to the original analysis (i.e. that the original second axis is now the first axis, etc.). Indeed, the results exactly correspond to the original analysis:

Axes					Total inertia
Eigenvalues :	0.172	0.120	0.101	0.096	5.311
Species-environment correlations :	0.831	0.881	0.880	0.911	
Cumulative percentage variance					
of species data :	3.5	6.0	8.1	10.0	
of species-environment relation:	15.3	26.0	35.0	43.5	

Sum of all eigenvalues		4.877
Sum of all canonical eigenvalues		1.126

Table 12-1. *Marginal and conditional effects obtained from the summary of forward selection*

Marginal effects			Conditional effects				
					Lambda		
Variable	Var.N	Lambda1	Variable	Var.N	A	P	F
Ca	1	0.35	Ca	1	0.35	0.002	4.79
Conduct	14	0.32	Conduct	14	0.13	0.002	1.79
Mg	2	0.30	Corg	12	0.11	0.002	1.60
pH	13	0.24	Na	5	0.12	0.002	1.58
Corg	12	0.24	NH_3	10	0.10	0.020	1.45
Na	5	0.18	Fe	3	0.09	0.018	1.34
NH_3	10	0.15	Cl	11	0.10	0.082	1.39
Si	6	0.12	pH	13	0.08	0.056	1.25
SO_4	7	0.12	Si	6	0.08	0.126	1.17
K	4	0.12	Mg	2	0.08	0.188	1.11
Fe	3	0.10	NO_3	9	0.08	0.332	1.07
Cl	11	0.10	SO_4	7	0.06	0.488	0.99
Slope	15	0.10	K	4	0.06	0.842	0.86
NO_3	9	0.09	PO_4	8	0.06	0.812	0.85
PO_4	8	0.08	Slope	15	0.06	0.814	0.83

The test of the first axis is significant ($F = 1.979$, $P = 0.018$). This shows that the original first axis, despite being clearly dominant, is not sufficient to explain the species–environment relationships in the data. In a similar manner, we could continue with testing the third and higher axes; but the third axis is not significant for this project.

You now know that there is a close correlation between environmental variables and species composition. You will use the forward selection to build a simpler model (with fewer explanatory variables), but one that still sufficiently explains the species composition patterns. First, it is useful to inspect the marginal effects of all environmental variables (i.e. the independent effect of each environmental variable). They can be easily obtained in CANOCO by asking for automatic forward selection, with the *best k* equal to the number of variables. You should ask for the Monte Carlo tests during the automatic forward selection, and you will also see the sequence of the forward selection steps, together with corresponding conditional effects (i.e. the effect that each variable brings in addition to all the variables already selected). Both tables can be obtained by using the '*FS-summary*' button in the Project View in Canoco for Windows. The results are shown in Table 12-1.

From the table with marginal effects, you can see that the calcium (*Ca*) concentration is the most important factor for species composition, followed by the *conduc*tivity, *Mg* concentration and *pH*. All these variables are closely correlated, as expected from their causal relationships – conductivity is in fact a measure of dissolved ions, and similarly pH is a function of the dissolved ion concentration. It is, consequently, not surprising that, after the Ca variable is selected, the conditional effects of conductivity, pH and Mg decrease dramatically. In fact, only the conductivity qualifies for the final model when the 0.05 probability threshold level for entry of a variable is adopted. Finally, the following variables are included in the model: Ca, conductivity, organic carbon (Corg), Na, NH_3 and Fe. The last two variables have relatively small marginal effects. However, they are independent of the other variables (note their low inflation factors in the Log View) and, because they probably affect the species composition, they add an explanatory power to the previously selected variables. You should note that the final selection tells you, in fact, that this is a sufficient set of predictors, and that further addition of variables does not significantly improve the fit. You should be extremely careful when trying to interpret the identified effects as the real causal relationships (e.g. by concluding that the effect of Mg is negligible because it is not included in the final selection). If variables in a group are closely correlated, then only a limited number of them is selected. It is often simply a matter of chance which of them is best correlated. After the best variable is selected, the conditional effects of variables correlated with it drop, sometimes dramatically. Even if they are functionally linked with the response (i.e. community composition), the test is rather weak. Also, the results of selection sometimes change with even a relatively minor change in the set of predictors from which we select the final model. However, this is a general problem of all observational studies with many correlated predictors.

The ordination diagram (Figure 12-2) reveals the same dominant gradient that was found in the previous analyses. However, the arch effect is less pronounced (without the use of detrending). Also, the effects of individual selected predictors can be better distinguished. When plotting the diagram, you can select the species either according to their weight or according to their fit to axes. But you should usually combine both methods – the species with highest weights are useful for comparison with the DCA, whereas the species with the highest fit are often ecologically more interesting. (Here, only the species with highest weight are shown in Figure 12-2, to enable comparison with the results of DCA in Figure 12-1.)

The diagram in Figure 12-2 shows the possible danger in selecting species according to the highest weight.[X] *Potentilla erecta*, *Festuca rubra*, *Carex nigra*,

---

[X] But similar problems can occur even when you combine fit and weight as selection criteria, because the species have to pass both limits to be included in the graph.

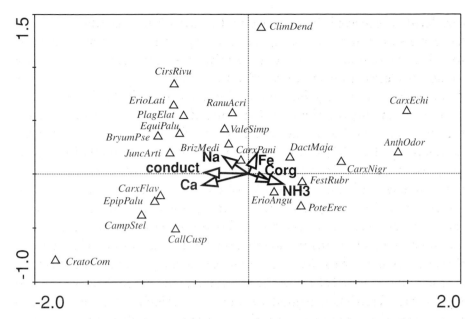

Figure 12-2. The species–environmental variables biplot of CCA with environmental variables selected by the forward selection procedure. The species with the highest weight are shown.

Anthoxanthum odoratum and _Carex echinata_ are suggested as most typical species for stands with a high soil carbon content; in fact, they are not very typical of such habitats. Because the acidic habitats were rare in comparison with the calcium-rich stands, their typical species were present in few samples, and consequently they did not have sufficient weight in the analysis. The selection according to the fit would display *Sphagnum* species to be typical for acidic bog habitats.

Now we can also inspect some synthetic community characteristics. In Figure 12-3, the size of the symbols corresponds to the species richness in individual samples.

To obtain this graph, select the *Create > Attribute Plots > Data Attribute Plots* command. In the *Attribute Plot Options* dialog, click in the *Select variable to plot* window on *Species > Sample Stats > Number of species*, and in the right side of the dialog select *Labeling: No labels, Visualization method: Symbol,* and *Additional content: Environmental variables*. You can see in the resulting diagram that the traditional wisdom of higher species diversity in calcium-rich habitats does not apply to this data set. When working with the number of species, one should be sure that the size of the sample plots was constant. In our case (as in other similar phytosociological surveys), the plot size changes slightly between samples. However, the noise that can be caused

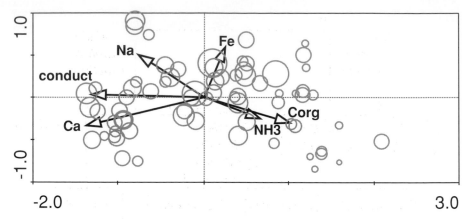

Figure 12-3. The sample–environmental variables biplot with symbol size corresponding to the number of species in the sample.

by variability in plot size is small relative to richness differences in the data.

12.3.   Classification

Another insight for this data set can be obtained from its classification. The TWINSPAN results are presented here as an example (Figure 12-4). The TWINSPAN was run with the default options (see Section 7.4), and the pseudospecies cut levels were 0, 2, 4 and 6. We will present the results only up to the second division (i.e. classification with four groups), although there were more divisions in the program output.

There are several possibilities as to how to use the environmental variables when interpreting classification (TWINSPAN) results. You can run, at each division, a stepwise discriminant analysis for the two groups defined by the division*; in this way, you will get a selection of environmental variables and a discriminant function, constructed as a linear combination of the selected environmental variables. The other possibility is to inspect, for each division, which environmental variables show the greatest differences between the groups (e.g. based on ordinary *t*-tests of all the variables; in such a heuristic procedure, you need not be too scrupulous about the normality). The environmental variables selected by the stepwise procedure will often not be those with the largest differences between the groups, because the stepwise selection is based on the largest conditional effects in each step, which means that

---

* The DISCRIM program doing exactly that is described in Jongman et al. (1987) and is available from the Microcomputer Power Co., USA, http://www.microcomputerpower.com

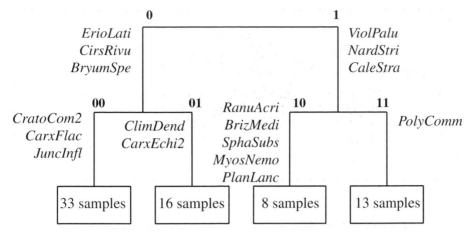

Figure 12-4. The results of the TWINSPAN classification. Each division is accompanied by the indicator species. More explanations are given in the text.

the predictors correlated with variables already selected are excluded. Both approaches, however, provide information that is potentially interesting. You can also compare the environmental values among the final groups only (as in Figure 12-5). But in this case, you will not obtain information about the importance of individual variables for individual divisions. Because these procedures involve data exploration rather than hypothesis testing, we are not concerned here about multiple testing and Bonferroni correction.

The search for a pattern is an iterative process, and is (at least in plant communities) usually based on the intimate biological knowledge of the species. Unlike hypothesis testing, at this stage the researcher uses his/her extensive field experience (i.e. the information that is external to the data set analysed). We have demonstrated here what can be inferred from multivariate data analyses. Further interpretations would be based on our knowledge of the biology of the species.

12.4.   Suggestions for additional analyses

There are many other analyses that one could do during the data exploration. Here we suggest just four interesting approaches:

1. The Ca and Mg ions are usually highly correlated in nature. We can be interested in the effect of one of them when the other one is kept constant. This can be achieved with two partial analyses: each of the variables will be a (single) environmental variable in one of the analyses and a (single) covariable in the other one.

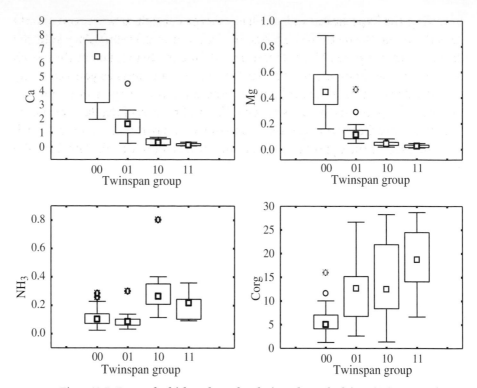

Figure 12-5. Box and whisker plots of molarity values of calcium (Ca), magnesium (Mg), NH_3 in the spring waters, and of the weight percentage of soil organic carbon, categorized according to the four groups defined by the TWINSPAN classification. Whiskers reach to the non-outlier extremes; outliers are shown as points.

2. We might be interested in how well correlated the species composition gradients are in vascular plants and in bryophytes. We can calculate the unconstrained ordination (DCA) first based on vascular plants only, and then on the bryophyte taxa only and then we can compare the sample scores on the ordination axes.

3. We might be interested in which of the environmental variables are the most important ones for vascular plants, and which for bryophytes. We can calculate separate constrained ordinations, and compare the importance of individual environmental variables (for example, by comparing models selected by the forward selection).

4. Further analyses are also possible with synthetic community characteristics, such as species richness, diversity and their relation to the environmental variables.

We have stressed several times that we are concentrating on a search for patterns in this case study, and not on hypotheses testing. If you are interested in testing the hypotheses, and you have a large enough data set, you should consider using the method suggested by Hallgren et al. (1999). They split the data set into two parts. The first one is used for data diving, i.e. for the search for patterns, without any restrictions (i.e. try as many tests as you want, any 'statistical fishing' is permitted). By this procedure, you will generate a limited number of hypotheses that are supported by the first part of the data set. Those hypotheses are then formally tested using the second part of the data set, which was not used in the hypothesis generation. See Hallgren et al. (1999) for the details.

Case study 3: Separating the effects of explanatory variables

13.1. Introduction

In many cases, the effects of several explanatory variables need to be separated, even when the explanatory variables are correlated. The example below comes from a field fertilization experiment (Pyšek and Lepš 1991). A barley field was fertilized with three nitrogen fertilizers (ammonium sulphate, calcium-ammonium nitrate and liquid urea) and two different total nitrogen doses. For practical reasons, the experiment was not established in a correct experimental design; i.e. plots are pseudo-replications. The experiment was designed by hydrologists to assess nutrient runoff and, consequently, smaller plots were not practical.[*] In 122 plots, the species composition of weeds was characterized by classical Braun–Blanquet relevés (for the calculations, the ordinal transformation was used, i.e. numbers 1–7 were used for grades of the Braun-Blanquet scale: $r, +, 1, \ldots, 5$). The percentage cover of barley was estimated in all relevés.

The authors expected the weed community to be influenced both directly by fertilizers and indirectly through the effect of crop competition. Based on the experimental manipulations, the overall fertilizer effect can be assessed. However, barley cover is highly correlated with fertilizer dose. As the cover of barley was not manipulated, there is no direct evidence of the effect of barley cover on the weed assemblages. However, the data enable us to partially separate the direct effects of fertilization from indirect effects of barley competition. This is done in a similar way to the separation of the effects of correlated predictors on the univariate response in multiple regression. The separation can be done using the variable of interest as an explanatory (environmental) variable and the other ones as covariables.

---

[*] We ignore this drawback in the following text.

Table 13-1. *The results of ANOVA on the complete regression model (calculated in the Statistica program)*

Analysis of Variance; DV: NSP (fertenv.sta)					
	Sums of Squares	DF	Mean Squares	F	p-level
Regress.	382.9215	2	191.4607	57.7362	2.98E-18
Residual	394.6195	119	3.31613		
Total	777.541				

13.2. Data

In this case study, you will work with simplified data, ignoring fertilizer type, and taking into account only the total fertilizer dose. Data are in the files *fertil.spe* (relevés) and *fertil.env* (fertilizer dose and barley cover) and also in the Excel file *fertil.xls* (so that you can try creating your own files for CANOCO using the WCanoImp utility). The dose of the fertilizer is *0* for unfertilized plots, *1* for 70 kg of N/ha, and *2* for 140 kg N/ha. The *Fertenv* worksheet also contains the number of species in each relevé. These data were also imported into the Statistica format (*fertenv.sta*). You will first do a univariate analysis of species numbers and then the corresponding analysis of species composition.

13.3. Data analysis

The univariate analysis will be demonstrated first. We have two predictors (explanatory, or 'independent' variables), *cover* and *dose*, and one response, number of species.[+] You will use multiple regression. In multiple regression, two tests are carried out. First, the complete model is tested by an analysis of variance. The null hypothesis states: **the response is independent of all the predictors**. The test results are in Table 13-1.

The null hypothesis was clearly rejected, and you can see that the regression sum of squares is roughly half of the total sum of squares (consequently, $R^2 = 0.5$), which means that this model explains half of the variability in the number of species. However, you still do not know which of the two explanatory variables is more important. This can be seen from the regression summary (Table 13-2).

The results show that the number of weed species decreases with both *cover* and *dose*, that the direct effect of *cover* is much more important, and also that

---

[+] This is also a simplification: one of the independent variables, *cover*, is dependent on the *dose* and consequently the use of path analysis might be a feasible solution here (see Legendre & Legendre 1998, p. 546). Nevertheless, from the point of view of the weed community, *dose* and *cover* can be considered as two correlated predictors.

Table 13-2. *The multiple regression summary, as calculated by the Statistica program*

	BETA	St. Err. of BETA	B	St. Err. of B	t(119)	p-level
Regression Summary for Dependent Variable: NSP (fertenv.sta)						
R = .70176746 R² = .49247757 Adjusted R² = .48394778						
F(2,119) = 57.736 p<.00000 Std.Error of estimate: 1.8210						
Intercpt			9.423662	0.388684	24.24506	0
DOSE	−0.02342	0.099678	−0.08501	0.361781	−0.23498	0.814629
COVER	−0.68390	0.099678	−0.06174	0.008999	−6.86113	3.28E-10

BETAs are the **standardized** partial regression coefficients (in effect, they are independent of the scale and consequently can be used to compare the importance of individual predictors), Bs are the usual (partial) regression coefficients, ts are the t-statistics for tests of the null hypotheses $B_i = 0$ (corresponding p is also presented; naturally, t and p values are the same for standardized and non-standardized coefficients).

cover is the only significant explanatory variable. If you calculate the univariate regressions on each of the predictors separately (i.e. their marginal effects), you will see that both of them are highly significant. This means that *dose* itself is a good predictor for the number of species, but does not significantly improve the fit when added to the predictor describing the cover of barley. On the other hand, *cover* is a good predictor for the number of species and significantly improves the fit when added to *dose*. We conclude that *cover* is sufficient to explain the species number.

We now proceed with the multivariate analysis.[$]

The following steps are recommended:

First, calculate an unconstrained ordination using detrended correspondence analysis (DCA). The results show the total variability by the length of the axes, which gives a measure of the total heterogeneity in the vegetation data. The length of the first axis in Hill's scaling is 3.8, which is in the 'grey zone' where both linear and unimodal methods should perform reasonably well.[*]

Second, calculate a constrained ordination with all of the environmental variables available. The length of the gradient from the first analysis serves as a lead for selecting between canonical correspondence analysis (CCA) and redundancy analysis (RDA). Another lead should be whether one can expect the majority of weed species to respond to fertilization and/or crop cover in a linear way or whether one can expect some (or all) species to have an optimum on this gradient. Generally, if the species change their proportions on the gradient then the linear approximation is correct. If you expect qualitative changes

---

[$] Note the difference: the species composition might change even when the number of species does not; in contrast, the test with the number of species might, in some cases, be stronger than the test with species composition. Consequently, the results might differ.

[*] See Section 4.3 for further details.

in the species composition (many species appearing and disappearing) then weighted averaging is better.

The authors of the published paper (Pyšek & Lepš 1991) used CCA. In this example, you will use RDA. RDA enables use of both the standardized and non-standardized analyses. Standardization by samples allows you to differentiate between the effect upon the total cover and upon the species composition. Use RDA (with the fertilizer *dose* and the barley *cover* as environmental variables) on data not standardized by sample norm. These results will reflect **both** the differences in total cover **and** the differences in the relative species composition. Also use RDA on data standardized by the sample norm where the results will reflect **only** the differences in relative representations of particular species.[+] In this case, the results of analyses with species data standardized by sample norm or non-standardized were rather similar (and were also similar to the output of CCA[#]). Only results of the non-standardized RDA will be shown in this tutorial.

For the third step, test the significance by a Monte Carlo permutation test, using unconstrained permutations. This is a test of the H_0: **there is no effect of the environmental variables on species representation**, i.e. the effect of both variables is zero. Because the analysis does not standardize the species values, even a proportional increase of all the species is considered a change in species composition. The rejection of the null hypothesis means that at least one of the variables has some effect on the species composition of the vegetation. The meaning of this test is analogous to an overall ANOVA of the multivariate regression model.

Next, inspect the analysis results. The effects of environmental variables are highly significant ($P = 0.002$ with 499 permutations – the highest possible with this number of permutations). The ordination diagram suggests the relationships of the species to the explanatory variables (Figure 13-1). As usual, the species with the highest fit with the axes are selected (*Project > Settings > Inclusion Rules*). The diagram shows that the explanatory variables are positively correlated, and most of the weed species are negatively correlated with both the dose and cover of barley. Only *Galium aparine*, a species able to climb stems of barley, is positively correlated.

There are other possibilities for displaying and exploring the results of the analysis provided by CanoDraw. In Figure 13-2, the sample scores based on the species data are displayed, with the size of each symbol corresponding to the number of species in the relevé. The environmental variables are also projected

[+] However, you should note that the interpretation of standardized data based on the species data values estimated on an ordinal scale may be problematic.

[#] This corresponds well to the fact that the length of the first DCA axis in turnover (SD) units indicated that both linear and unimodal methods would perform well.

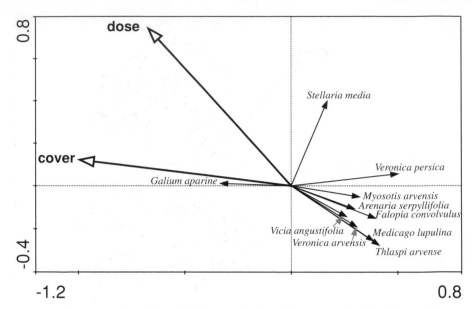

Figure 13-1. The ordination diagram of RDA with non-standardized data, with two explanatory variables, *cover* and *dose*.

onto the diagram. The diagram displays clearly how the species richness decreases with both dose and cover.

In this figure, the sample scores that are based on species compositions (i.e. sample scores that are linear combinations of species in the sample, the *Samp* scores in a CANOCO solution file) are used. In CanoDraw, the default for constrained ordination is to use the sample scores that are linear combinations of the environmental values (*SamE* scores in a CANOCO solution file). Although the constrained ordination is constructed to minimize the difference between the *Samp* and *SamE* scores, each of the two sets provides different information. *SamE* scores are the values fitted by the model; in other words these scores indicate where the samples should be according to the fitted model. As a result, when the *SamE* scores are displayed, all the samples with the same values of environmental variables (i.e. all the samples with the same treatment in this example) have exactly the same position in the ordination diagram.

Here it might be better to use the *Samp* scores to display the variability in the species composition. To do this in CanoDraw, select *Project* > *Settings* > *Contents* and check the *Plot SAMP scores even for . . .* option. To create the diagram from Figure 13-2, select the *Create* > *Attribute Plots* > *Data Attribute Plot* command, select the *Species* > *Sample Stats* > *Number of species* in the listbox and select (or check) the *No labels, Symbols* and *Environmental variables* options on the right side of the dialog.

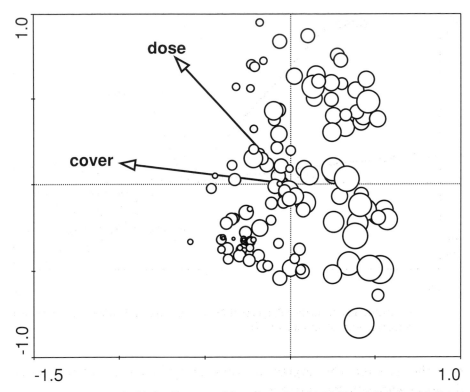

Figure 13-2. The biplot diagram with environmental variables and sites, with the size of the site symbols corresponding to species richness (number of species in relevé).

The same information can also be presented by smoothed data (particularly when you have many samples). To do this, select again the *Create > Attribute Plots > Data Attribute Plot* command. In the dialog, select again *Species > Sample Stats > Number of species*, and in the *Visualization Method* area an appropriate method for smoothing (e.g. *Loess*). Do not forget to check the *Environmental variables* option in the *Additional contents* area. After you click the *OK* button, you will be asked to specify details of the smoothing method. The plot in Figure 13-3 was obtained by using the default values for the loess method.

Finally, it is necessary to calculate two separate partial constrained ordinations. In the first case, use the fertilizer *dose* as the environmental variable and the barley *cover* as the covariable. The analysis will reflect the effect of fertilization, which cannot be explained by the effect caused by the increased barley cover (the conditional effect of fertilizer).

In the second analysis, use barley *cover* as the environmental variable and fertilizer *dose* as the covariable. Here you ask about the conditional effect of barley cover. This analysis reflects the variability caused by the differences in

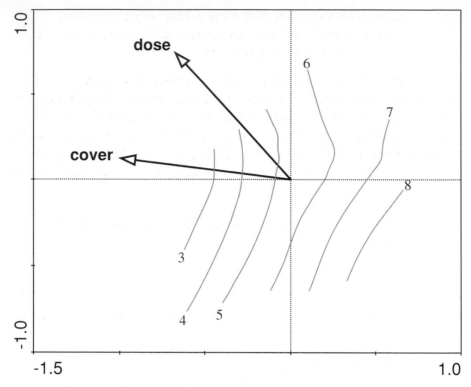

Figure 13-3. The isolines of species richness of samples, plotted in the RDA ordination diagram.

barley cover that cannot be attributed to fertilization effects. Use both the analyses standardized and not standardized by sample norm. It is possible to use constrained permutations (permutations within blocks) in the Monte Carlo permutation test when dose is used as a covariable. Because dose has three separate values, the plots with the same value of the dose can be considered to form a block. However, with the reduced model, use of within-block permutations is not necessary, and unconstrained permutations might provide a stronger test. The meaning of these tests of conditional effects is analogous to a test of significance of particular regression coefficients in multiple regression.

> Note that it may sometimes happen (not in this data set) that the analysis with both variables used as explanatory variables is (highly) significant, whereas none of the analyses with one of the variables used as an explanatory variable, and the other as a covariable, is significant. This case is equivalent to the situation in a multiple linear regression where the ANOVA on the total regression model is significant and none of the regression coefficients differ significantly from zero. This happens when the predictors are highly correlated.

You are then able to say that the predictors together explain a significant portion of total variability, but you are not able to say which of them is the important one.

Variation partitioning (see Section 5.10 for additional information) decomposes the total variability into a part that can be explained solely by the dose and a part solely explained by the cover. As the two variables are highly correlated, it is impossible to say, for some part of the variability, which of the variables causes it. To calculate the individual parts of the variability decomposition, you can use the results of the redundancy analyses that you have already calculated:

1. The RDA with both variables used as explanatory (environmental) variables. You get:

Axes		1	2	3	4	Total variance
Eigenvalues	:	0.097	0.046	0.200	0.131	1.000
Species-environment correlations	:	0.705	0.513	0.000	0.000	
Cumulative percentage variance						
Cumulative percentage variance of species data	:	9.7	14.3	34.4	47.5	
of species-environment relation	:	67.8	100.0	0.0	0.0	
Sum of all eigenvalues						1.000
Sum of all canonical eigenvalues						**0.143**

The results show that 14.3% of the total variation in species data can be explained by both variables together.

2. Now use the partial analysis with cover as the explanatory (environmental) variable and dose as a covariable:

Axes		1	2	3	4	Total variance
Eigenvalues	:	0.074	0.200	0.131	0.102	1.000
Species-environment correlations	:	0.590	0.000	0.000	0.000	
Cumulative percentage variance of species data	:	8.0	29.5	43.6	54.5	
of species-environment relation	:	100.0	0.0	0.0	0.0	
Sum of all eigenvalues						0.931
Sum of all canonical eigenvalues						**0.074**

The sum of all eigenvalues is after fitting covariables
Percentages are taken with respect to residual variances
        i.e. variances after fitting covariables

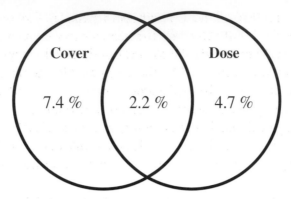

Figure 13-4. Partitioning of variance of species composition, explained by the *cover* and *dose* factors.

This shows that cover explains 7.4% of the total variability in the species data, which cannot be explained by dose. Note that you do not use the percentage presented in the cumulative percentage variance of species data (8%), because it is taken with respect to the residual variation (i.e. calculated after fitting the covariable), but you directly use the appropriate canonical eigenvalue.

3. In a similar manner, you can now use the partial ordination with dose as the explanatory variable and cover as a covariable:

```
                                                             Total
Axes                                    1     2     3     4   variance

Eigenvalues                        : 0.047 0.200 0.131 0.102 1.000
Species-environment
   correlations                    : 0.500 0.000 0.000 0.000
Cumulative percentage variance
   of species data                 :   5.2  27.4  41.9  53.2
   of species-environment
   relation                        :   0.0   0.0   0.0
Sum of all          eigenvalues                              0.904
Sum of all canonical eigenvalues                             0.047
```

The dose explains 4.7% of variability not accounted for by the cover.

4. Now you can calculate: total explained variability is 14.3%. Of those, 7.4% can be explained solely by the effect of barley cover and 4.7% by the effect of fertilizer dose. It is impossible to decide which of the environmental variables is responsible for the remaining 2.2% of variability. The results of variation partitioning can be displayed in a diagram (Figure 13-4).

Do not be discouraged that the explained variability is only 14.3%! The explained variability also depends on the number of species and sites (on the dimensionality of the problem). In the RDA with both variables, the third (i.e. unconstrained) axis explains only 20%, and the first axis in a corresponding PCA explains only 23.9%. The first axis of PCA can be understood as expressing how much the data could be explained by the best possible environmental variable. From this point of view, the explanatory power of cover and dose is not so bad. When evaluating the results of the variance partitioning procedure, you should always compare the explained amounts of variance with the results of unconstrained analyses. Nevertheless, we do not think that you should really ignore the unexplained fraction of variability, as some authors suggest (e.g. Okland 1999).

When you compare the results obtained here with those of Pyšek & Lepš (1991), you will find important differences. This is caused by the omission of fertilizer type as an explanatory variable in your sample analyses. There were relatively large differences in cover among the fertilizer types within a dose. All these differences are now ascribed to the effect of barley cover. This also illustrates how biased the results can be when some of the important predictors in the data are ignored in a situation when predictors are not experimentally manipulated and are interdependent.

14

Case study 4: Evaluation of experiments in randomized complete blocks

14.1. Introduction

Randomized complete blocks design is probably the most popular experimental design in ecological studies, because it controls in a powerful way the environmental heterogeneity. For a univariate response (e.g. number of species, total biomass) the results of experiments set in randomized complete blocks are evaluated using a two-way ANOVA without interactions. The interaction mean square is used as the error term – the denominator in the calculation of F-statistics. In the following tutorial, you will use the program CANOCO in a similar way to evaluate the community response (i.e. a multivariate response of the species composition of the vegetation). The example is based on an experiment studying the effect of dominant species, plant litter and moss on the composition of a community of vascular plants, with special attention paid to seedling recruitment. In this way, some of the aspects of the importance of regeneration niche for species coexistence were tested. The experiment was established in four randomized complete blocks, the treatment had four levels and the response was measured once. The experiment is described in full by Špačková et al. (1998). Here is a simplified description of the experiment.

The experiment was established in March 1994, shortly after snowmelt, in four randomized complete blocks. Each block contained four plots, each with a different treatment: (1) a control plot where the vegetation remained undisturbed; (2) a plot with the removal of litter; (3) a plot with removal of the dominant species *Nardus stricta*; and (4) a plot with removal of litter and mosses.* Each plot was 2 m × 2 m square. The original cover of *Nardus stricta* was about

---

* It seems that one treatment is missing – removal of mosses only; however, for practical reasons it was impossible to remove mosses without removing the litter.

25%. Its removal was very successful, with nearly no re-growth. The removal in the spring caused only minor soil disturbance that was not apparent in the summer.

14.2.   Data

In each central 1 m² plot, cover of adult plants and bryophytes was visually estimated in August 1994. At that time, a new square (0.5 m × 0.5 m) was marked out in the centre of each plot and divided into 25 0.1 m × 0.1 m subplots. In each subplot adult plant cover and numbers of seedlings were recorded. In this case study, you will use the seedling totals in the 0.5 m × 0.5 m plots. The data are in the CANOCO-format files *seedl.spe* (species data) and *seedl.env* (design of the experiment). In the *seedlenv.sta* file, the data on total number of seedlings in quadrats are also presented. Those will be used to demonstrate the analogous univariate analysis. Original data are in the *seedl.xls* spreadsheet file.

14.3.   Data analysis

You may ask whether it is necessary to use a multivariate method. Would it not be better to test the effect on each species separately by a univariate method (i.e. either by ANOVA, or by analysis of deviance)? It is much better to use a multivariate method. There is a danger in using many univariate tests. If you perform several tests at the nominal significance level $\alpha = 0.05$, the probability of a type I error is 0.05 **in each univariate test.** This means that when testing, say, 40 species, you could expect two significant outcomes just as a result of type I error. This can lead to 'statistical fishing', when one picks up just the results of significant tests and tries to interpret them. We could overcome this by using the Bonferroni correction,[*] but this leads to an extremely weak test. We consider it possible (but some other statisticians probably would not) to use the univariate methods for particular species when we find the community response significant; however, you should keep in mind that the probability of a type I error is α in each separate test. You should be aware that if you select the most responsive species according to the results of ordination,

---

[*] **Bonferroni correction** means, in this context, dividing the nominal significance level by the number of tests performed and performing the particular tests on the resulting significance level. This assures that the overall probability of the type I error in at least one of the tests is equal or smaller than α (see Rice 1989 or Cabin & Mitchell 2000).

Table 14-1. *Results of ANOVA on randomized complete blocks design*

Summary of all Effects; design: (seedlenv.sta)					
1-TREATMEN, 2-BLOCK					
	Df Effect	MS Effect	df Error	MS Error	F	p-level
1	3	4513.229	9	1068.84	4.222548	0.040278
2	3	215.5625	9	1068.84	0.201679	0.892645
12						

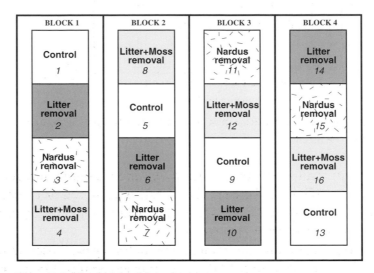

Figure 14-1. The design of the experiment.

those species very probably will give significant differences, and the univariate test does not provide any further information in addition to the results of ordination. Nevertheless, even when there are no differences between the treatments at all, **in your data set** some species will be more common in some of the treatments, just by chance. When you then plot the ordination diagram, select the 'most responsive' species (by eye) and test them by ANOVA, the result will very probably be significant.

The design of the experiment is shown in Figure 14-1. Each quadrat is characterized by (1) the type of the treatment, and (2) the block number. Performing the univariate analysis is quite simple. In the *ANOVA/MANOVA* module of the Statistica program you declare *TREATMEN*t and *BLOCK* to be the independent variables and *SEEDLSUM* (the total number of seedlings in a quadrat) to be the dependent variable. In randomized complete blocks, the interaction between treatment and blocks is used as the error term. State this in the next panel (by using *Pooled effect/error term > Error*, and then selecting *TREATMEN*BLOCK*). You will get the ANOVA table (Table 14-1).

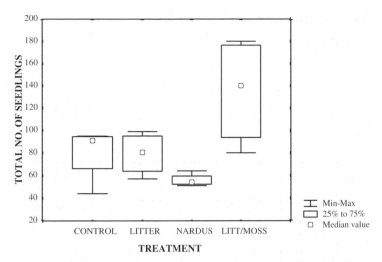

Figure 14-2. Box-and-whisker plot of number of seedlings in individual treatments.

The analysis results show that the treatment effect is significant, but there are hardly any differences between the blocks. Because the block does not explain any variability, the analysis is weaker than it would be if you disregarded the block effects (then, $p = 0.015$).

Now, you can use the multiple comparisons for testing pair-wise differences (a little tricky in Statistica for ANOVA in randomized complete blocks), and/or inspect the box-and-whisker plot (Figure 14-2). You can see that only the removal of both litter and moss has a significant effect on the total seedling number.*

For the analysis with CANOCO, you should note that both treatment and block are categorical variables and consequently have to be coded as a set of 'dummy' variables. The variables characterizing the block structure will be used as covariables and the treatment variables will be the environmental variables. The corresponding data set is displayed in Table 14-2. CANOCO asks separately for the environmental data and for the covariables, so each of those can be in a separate file. If all the environmental variables and covariables are in the same file, you have to specify the same file name (*seedl.env* in this case) for both the environmental variables and the covariables and then omit the treatment variables from the covariable list and the block variables from the list of environmental variables. It is very convenient to have the environmental variables

---

* Using the Duncan test, the combined removal treatment is different from all the other treatments, while Tukey's test, which is usually recommended, shows that the treatment differs only from the *Nardus* removal.

Table 14-2. *Environmental data characterizing the design of the experiment (CANOCO file in full format)*

```
Ohrazeni 1994-seedlings, design of experiment
  (i3,8(f3.0))
  8
  1   1   0   0   0   1   0   0   0
  2   0   1   0   0   1   0   0   0
  3   0   0   1   0   1   0   0   0
  4   0   0   0   1   1   0   0   0
  5   1   0   0   0   0   1   0   0
  6   0   1   0   0   0   1   0   0
  7   0   0   1   0   0   1   0   0
  8   0   0   0   1   0   1   0   0
  9   1   0   0   0   0   0   1   0
 10   0   1   0   0   0   0   1   0
 11   0   0   1   0   0   0   1   0
 12   0   0   0   1   0   0   1   0
 13   1   0   0   0   0   0   0   1
 14   0   1   0   0   0   0   0   1
 15   0   0   1   0   0   0   0   1
 16   0   0   0   1   0   0   0   1

  control  litter—rnardus—rlit+mossblock1  block2  block3  block4
  rel1     rel2     rel3      rel4     rel5    rel6    rel7    rel8 etc.
  rel11    rel12    rel13     rel14    rel15   rel16
```

and the covariables in the same file. You can select various subsets of environmental variables and covariables in various partial analyses.

The fourth environmental variable describing the last type of treatment (the last level of the factor) is not necessary here: three 'dummy' variables are sufficient for coding one categorical variable with four categories. The fourth variable will be omitted from the calculations, as it is a linear combination of the previous three variables ('lit + moss' = 1 − 'control' − 'litter-r' − 'nardus-r'). However, it is useful to include it in the analysis because we will need to plot the results using CanoDraw after the calculations. In this case, you would like to plot the centroids of all four categories. Similarly, the separate (co-)variable for the fourth block is not necessary.

As the vegetation in the plots is very homogeneous and you also have only categorical explanatory variables, you will use **redundancy analysis** (RDA), the method based on the linear model. Also, the length of the gradient on the first axis of a DCA is 1.98; on the second axis, 1.41.* Even more importantly, RDA (unlike CCA) also enables you to carry out both the

---

* See Section 4.3 for explanation.

standardized and non-standardized analyses. Now, you can test (at least) two hypotheses:

1. The first null hypothesis can be stated as follows: **there is no effect of the manipulation on the seedlings.** To reject this hypothesis, it is enough if the total number of seedlings differs between treatments, even if the proportion of individual seedling species remains constant. When the proportion of seedlings changes or when both the proportion and the total number of seedlings change, the null hypothesis will obviously also be rejected.

2. The second null hypothesis can be stated as: **the relative proportions of species among the seedlings do not differ between the treatments.** Rejecting this hypothesis means that the seedlings of different species differ in their reaction to the treatment. The first hypothesis can be tested only when you use no standardization by samples (the default in CANOCO). When you use a standardization by samples (usually by the sample norm), then you test the second null hypothesis.* The test of the first hypothesis above is usually more powerful, but the rejection of the second hypothesis is more ecologically interesting: the fact that seedlings of different species respond in different ways to particular treatments is a good argument for the importance of the regeneration niche for maintenance of species diversity.

The calculation of RDA proceeds in a standard way. Just take care of the following issues:

1. When you have environmental variables and covariables in the same file do not forget to omit the appropriate variables. Omitting of covariables is more important, because if the same variable is among the covariables *and* the environmental variables, it will be automatically omitted from the environmental variables, as it does not explain any variability.

2. When performing the Monte Carlo permutation test, you can ask for permutation within blocks, conditioned by all three covariables (the fourth is collinear and is omitted from the analyses). Each permutation class will then correspond to a block in the experiment. The permutation within blocks is shown in Table 14-3. This is the approach recommended in the versions of CANOCO before 4.x (under the null hypothesis, the treatments are freely exchangeable within a block). In the newer versions,

---

* Note that the standardization is part of the averaging algorithm in canonical correspondence analysis. Consequently, a CCA would not find any difference between the plots differing in total number of seedlings, but with constant proportions of individual species.

Table 14-3. *Permutations within blocks*

original	block	perm1	perm2	perm3	perm 4	perm 5
1	1	2	4	1	etc.	
2	1	4	3	4	etc.	
3	1	3	2	2	etc.	
4	1	1	1	3	etc.	
5	2	7	5	7	etc.	
6	2	8	8	6	etc.	
7	2	5	7	8	etc.	
8	2	6	6	5	etc.	
9	3	11	9	9	etc.	
10	3	9	12	12	etc.	
11	3	10	10	11	etc.	
12	3	12	11	10	etc.	
13	4	14	16	14	etc.	
14	4	15	13	15	etc.	
15	4	16	15	13	etc.	
16	4	13	14	16	etc.	

when using the reduced model test, the residuals after subtracting the effect of covariables are permuted. In this case, we recommend using unrestricted permutations (the increased strength is achieved without inflating the type I error rate) (Anderson & Ter Braak 2002). The test based on the within-block permutations is particularly weak with a low number of blocks and/or a low number of treatment levels.

Of the two suggested analyses, only the RDA on data not standardized by sample norm shows significant results ($P = 0.028$ for the test on the first axis, $P = 0.056$ for the test on all constrained axes, using 499 unrestricted permutations in each test). The fact that the test of the first axis is much stronger suggests that there is a strong univariate trend in the data. Accordingly, the second canonical axis is not significant (see Section 9.1 for a description of testing the significance of higher axes). Also, the ordination diagram (species–environment biplot, Figure 14-3) confirms that the most different treatment is the removal of moss and litter, with the samples with this treatment separated from the others on the first canonical axis.

In Log View, you should notice the striking difference between the eigenvalues corresponding to the first and to the second ordination axes (0.281 and 0.031, respectively). This again suggests that the explanatory power of the second axis is very weak. Also, with this difference between eigenvalues, there is a huge difference between graphs produced with scaling focused on species correlation and on the inter-sample distances. As the centroids of treatments are calculated as averages of individual sample positions, the scaling focused

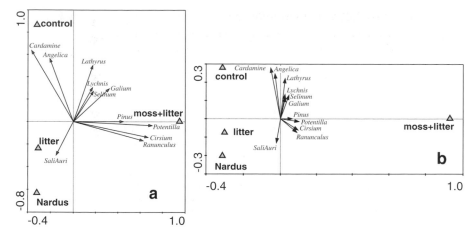

Figure 14-3. Species–environment biplot of RDA on non-standardized data. Species are shown by arrows and labelled by their generic names only, and the centroids of treatments are displayed by triangles. Both diagrams show the scores on the first two axes of identical analysis, but the left one (a) was produced with scaling of scores focused on inter-species correlations and the right one (b) with scaling focused on inter-sample distances.

on species correlations exaggerates the differences between treatments on the second axis (compare the two diagrams in Figure 14-3).

If you needed to make multiple comparisons, you would have to calculate separate analyses with pairs of treatments (omitting all the samples with the other treatments than the two compared; this change can be performed during the project setup), and then use the Bonferroni correction (in effect to multiply each *P* value obtained in a particular test by the number of tests executed). This gives, however, a very weak test.

When plotting the ordination diagram, the species–environment biplot is the most informative display. In CanoDraw, two things should not be forgotten: (1) specify the experimental treatments as nominal variables (*Project > Nominal variables > Environmental variables*) and (2) restrict the number of species in the final graph. The species can be selected manually, but more useful is the selection of species with the highest fit. Use *Project > Settings > Inclusion Rules* and modify (by the trial-and-error method) the lower limit of the *Species Fit Range* option to achieve a reasonable number of species in the final plot.

The ordination diagram also shows that *Ranunculus, Potentilla* and *Cirsium* are the species responsible for differentiation and all of them prefer the treatment with the removal of moss and litter. Although it seems that particular species prefer particular treatments, the standardized RDA was not significant

($P \sim 0.5$), so you are not able to confirm that there are differences between species in their seedlings' reaction to the treatments.

There is an important difference between the ecological implications of significant treatment effect in the standardized-by-samples and the non-standardized analyses. The theory of a regeneration niche (Grubb 1977) suggests that differential establishment of seedlings of various species in various microhabitats is a mechanism that supports species coexistence. The described experiment tests this suggestion. If there was differential establishment, then we could expect the proportions of species to be different in different treatments. Unfortunately, we were not able to demonstrate this. We were only able to show that mosses and litter suppress seedling recruitment, but this conclusion does not directly support Grubb's theory.

In this experiment, there were hardly any differences between the blocks ($P \sim 0.5$). Consequently, when you ignore the blocks, both in ANOVA and in RDA, you will get stronger tests. This is a nice illustration of the fact that the blocks that do not explain any variability decrease the power of the tests (for the non-standardized RDA, P would be 0.015 and 0.036 for the tests on the first axis and on the sum of canonical eigenvalues, respectively).

Case study 5: Analysis of repeated observations of species composition from a factorial experiment

15.1. Introduction

Repeated observations of experimental units are typical in many areas of ecological research. In experiments, the units (plots) are sampled first **before** the experimental treatment is imposed on some of them. In this way, you obtain 'baseline' data, i.e. data where differences between the sampling units are caused solely by random variability. After the treatment is imposed, the units are sampled once or several times to reveal the difference between the **development** (dynamics) of experimental and control units. This design is sometimes called replicated BACI – before after control impact. To analyse the univariate response (e.g. number of species, or total biomass) in this design, you can usually apply the repeated measurements model of ANOVA.

There are two possibilities for analysing such data. You can use the split-plot ANOVA with time, i.e. the repeated measures factor, being the 'within plot' factor,* or you can analyse the data using MANOVA. Although the theoretical distinction between those two is complicated, the first option is usually used because it provides a stronger test. Nevertheless, it also has stronger assumptions for its validity (see e.g. Lindsey 1993; Von Ende 1993). The interaction between time and the treatment reflects the difference in the development of the units between treatments. CANOCO can analyse the repeated observations of the species composition in a way equivalent to the univariate repeated measurements ANOVA. Whereas all the model terms are tested simultaneously in ANOVA, in CANOCO you must run a separate analysis to test each of the terms. We will illustrate the approach with an analysis of a factorial experiment

---

* This is called a 'univariate repeated measurements ANOVA'.

applying fertilization, mowing and dominant removal to an oligotrophic wet meadow. The description of the experiment is simplified; a full description is in Lepš (1999).

15.2.   Experimental design

In this experiment, we tested the reaction of a plant community to various management regimes and their combinations (mowing, fertilization), and the effect of the dominant species tested by its removal. We were interested in the temporal dynamics of species richness and species composition under various treatments, and also in which species traits are important for their reaction.

The experiment was established using a factorial design with three replications of each combination of treatments in 1994. The treatments were fertilization, mowing and removal of the dominant species (*Molinia caerulea*). This implies eight combinations in three replications, yielding 24 2m × 2m plots. The fertilization treatment is an annual application of 65 g/m² of commercial NPK fertilizer. The mowing treatment is the annual scything of the quadrats in late June or early July. The cut biomass is removed after mowing. *Molinia caerulea* was manually removed (using a screwdriver) in April 1995 with a minimum of soil disturbance. New individuals are removed annually.

15.3.   Sampling

Plots were sampled in the growing season (June or July) each year, starting in 1994. Note that initial sampling was conducted before the first experimental manipulation in order to have baseline data for each plot. The cover of all vascular species and mosses was visually estimated in the central 1 m² of the 2 m × 2 m plot.

15.4.   Data analysis

Data are in the form of repeated measurements; the same plot was sampled four times. For a univariate characteristic (the number of species) the corresponding repeated measurements ANOVA model was used (Von Ende 1993). For species composition, RDA was used: RDA, a method based on a linear species response, was used because the species composition in the plots was rather homogeneous and the explanatory variables were categorical. Because *Molinia* cover was manipulated, **this species was specified as a supplementary species in the analyses. This is very important because otherwise** we would show (with high significance) that *Molinia* **has a higher cover in the plots from which it was not removed**. By using the various combinations of explanatory ('environmental' in CANOCO terminology) variables and covariables in RDA with the appropriate permutation scheme in the Monte Carlo test, we were able to construct tests analogous to the testing of significance of particular terms in ANOVA models (including repeated measurements). Because the data form repeated observations that include the baseline (before treatment) measurements, the **interaction of treatment and time** is of the greatest interest and corresponds to the effect of the experimental manipulation. When we test for the interaction, the plot identifiers (coded as many dummy variables) are used as covariables. In this way we subtract the average (over years) of each plot, and only the changes within each plot are analysed. Values of time were 0, 1, 2 and 3 for the years 1994, 1995, 1996 and 1997, respectively. This corresponds to a model where the plots of various treatments do not differ in 1994 and the linear increase in difference is fitted by the analysis.*

The other possibility is to consider time as a categorical (i.e. nominal) variable (each year would be a separate category) and to code it as several dummy variables. In a 'classical' analysis using constrained ordination and diagrams, both approaches can be used (but only the analyses using time as a quantitative variable will be shown in this tutorial). A novel method of visualization of results of repeated measurements analysis, the principal response curves method (PRC, Van den Brink & Ter Braak 1999) has been suggested. It is an extension of constrained ordinations and time is used as a categorical variable. In this chapter, we will demonstrate the univariate analysis, the classical constrained ordinations and the PRC method. Because the creation of PRC curves is more tricky than the other topics, we present it in a more detailed, cookbook fashion.

The original data are in Excel file *ohraz.xls*. The species data are in the worksheet *ohrazspe*, and the design of the experiment is in the *ohrazenv* worksheet. Using the program WCanoImp, prepare condensed-format CANOCO species file *ohraz.spe* and environmental data file *ohraz.env*. In both files, the samples are in the following order: samples from 1994 have numbers 1 to 24, samples from 1995 are numbered 25 to 48, etc. The knowledge of this ordering will be important for the description of the permutation scheme. The names of samples are constructed in the following way: r94p1 means a sample recorded in 1994, plot 1. In the environmental data file, the first three variables (*MOWING*,

---

* This approach is analogous to the linear polynomial contrasts rather than the ordinary testing of effects in repeated measurement ANOVA.

FERTIL, REMOV) describe which treatments were applied using the following values: 1 = treated, 0 = non-treated. The next variable, *Year*, is the time from the start of experiment, i.e. time as a quantitative variable. The next four variables (*Yr0, Yr1, Yr2* and *Yr3*) are dummy variables, describing the sampling year as a categorical variable (so for all the records done in 1994, Yr0 = 1, and Yr1 = 0, Yr2 = 0 and Yr3 = 0). The following variables, *P* 1 to *P* 24, are the plot identifiers (e.g. for all the records from plot one, P1 = 1 and P2 to P24 are zero). You will use this file as both environmental variables and covariables and select (delete) appropriate variables in each context.

15.5.   Univariate analyses

At first, the univariate analysis of the data on number of species in quadrats will be demonstrated using the *ANOVA/MANOVA* module in the Statistica program. The data on species numbers are in the worksheet *ohrstat* and also in the Statistica file *ohrazenv.sta*. Note that unlike for the multivariate analysis, where each relevé has to be presented as a separate sample, here each plot is a separate sample, characterized by four variables (i.e. the repeated measurements of the species richness in the four consecutive years). The data on species richness will be analysed by a repeated measurement ANOVA.

In Statistica, you first state that the independent variables are MOWING, FERTILization and REMOVal, and the dependent variables are numbers of species in the four consecutive years (i.e. *NSP0, NSP1, NSP2* and *NSP3*). Then you specify that the repeated measure design variable has four levels, and its name is *TIME*. After confirming the information and asking for all the effects, you will get the ANOVA table (Table 15-1).

Together, 15 various tests were performed.* Of greatest interest here are the interactions of treatments with time highlighted in Table 15-1 (and all of them are significant). The main effects and their interactions are also of interest, because sometimes they may provide a stronger test than the interaction.

15.6.   Constrained ordinations

For most of the tests shown in Table 15-1 it is possible to construct a counterpart in the CANOCO analysis by a combination of environmental variables and covariables.

---

* Interestingly, when you perform several *t*-tests instead of multiple comparisons controlling the experiment-wise type I error rate, you will get a clear response from some journal reviewers; when one performs 15 tests in a multiway ANOVA, the type I error rate is also given for each individual test, and nobody cares. (We realize that there **is** a difference between these two situations, but both provide an opportunity for 'statistical fishing'.)

Table 15-1. *Results of univariate repeated measurements analysis of variance from the Statistica program*

```
Summary of all Effects; design: (ohrazenv.sta)
1-MOWING, 2-FERTIL, 3-REMOV, 4-TIME
```

	df Effect	MS Effect	df Error	MS Error	F	p-level
1	1	65.01041	16	40.83333	1.592092	0.225112
2	1	404.2604	16	40.83333	9.900255	0.006241
3	1	114.8438	16	40.83333	2.8125	0.112957
4	3	87.95486	48	7.430555	11.83692	6.35E-06
12	1	0.260417	16	40.83333	0.006378	0.937339
13	1	213.0104	16	40.83333	5.216582	0.036372
23	1	75.26041	16	40.83333	1.843112	0.193427
14	**3**	**75.53819**	**48**	**7.430555**	**10.16589**	**2.69E-05**
24	**3**	**174.2882**	**48**	**7.430555**	**23.45561**	**1.72E-09**
34	**3**	**41.48264**	**48**	**7.430555**	**5.58271**	**0.002286**
123	1	6.510417	16	40.83333	0.159439	0.694953
124	3	14.67708	48	7.430555	1.975234	0.130239
134	3	11.48264	48	7.430555	1.545327	0.214901
234	3	2.565972	48	7.430555	0.345327	0.792657
1234	3	3.538194	48	7.430555	0.476168	0.700348

Interactions of treatments with time are displayed in bold typeface.

Table 15-2. *Results of the analyses using RDA applied to cover estimates of the species in 1m × 1m plots*

Analysis	Explanatory variables	Covariables	% expl. 1st axis	r 1st axis	F ratio	P
C1	Yr, Yr*M, Yr*F, Yr*R	PlotID	16.0	0.862	5.38	0.002
C2	Yr*M, Yr*F, Yr*R	Yr, PlotID	7.0	0.834	2.76	0.002
C3	Yr*F	Yr, Yr*M, Yr*R, PlotID	6.1	0.824	4.40	0.002
C4	Yr*M	Yr, Yr*F, Yr*R, PlotID	3.5	0.683	2.50	0.002
C5	Yr*R	Yr, Yr*M, Yr*F, PlotID	2.0	0.458	1.37	0.084

Data are centred by species. No standardization by samples was done. Explanatory variables are environmental variables in CANOCO terminology. *% expl. 1st axis*: percentage of species variability explained by the first ordination axis, a measure of the explanatory power of the variables. *r 1st axis*: species–environment correlation on the first axis. *F ratio*: the *F*-ratio statistics for the test on the trace. *P*: corresponding probability value obtained by the Monte Carlo permutation test, using 499 random permutations. *Yr*: serial number of the year; *M*: mowing; *F*: fertilization, *R*: *Molinia* removal, *PlotID*: identifier of each plot. The asterisk (*) between two terms indicates their interaction.

We selected to test the following null hypotheses (see Table 15-2):

- *C1*: there are no directional changes in time in the species composition that are common to all the treatments or specific for particular treatments. This corresponds to the test of all within-subject effects in a repeated measures ANOVA.

- *C2*: The temporal trend in the species composition is independent of the treatments.
- *C3* (*C4*, *C5*): Fertilization (or removal or mowing) has no effect on the temporal changes in the species composition. This corresponds to the tests of particular terms in a repeated measures ANOVA (the three highlighted rows in Table 15-1).

Note that when *PlotID* is used as a covariable, the main effects (i.e. *M*, *F* and *R*) do not explain any variability and it is meaningless to use them, either as covariables or as the explanatory variables.

In these analyses, time is considered as a quantitative (continuous) variable. This means that you will use *Year* and delete the four variables *Yr0* to *Yr3*. This corresponds to a search for a linear trend in the data. If you look for general differences in the temporal dynamics, you can consider time to be a categorical variable and use the variables *Yr0* to *Yr3* (one of them is redundant, but useful for plotting in the ordination diagrams) and delete the variable *Year*. In this case, however, the interaction between treatment and time must be defined as interactions with all four dummy variables *Yr0* to *Yr3*.

During the project setup in Canoco for Windows, you should not forget to tick the checkboxes *Delete species* (and delete *Molinia*; alternatively, you can make this species supplementary), *Delete environmental variables*, *Delete covariables* and *Define interactions* of environmental variables and/or covariables, where necessary. When defining an interaction, you can use the effects that were deleted from the variables. So, for testing hypotheses *C2* to *C5* all the environmental variables should be deleted and then their interactions defined.

The null hypothesis *C1* is a little more complicated and difficult to interpret ecologically. This analysis is useful for a comparison with the other analyses in terms of the explained variability and of the species – environment correlation on the first axis. For the other analyses, under the null hypotheses, the dynamics are independent of the treatments imposed. This means that if the null hypothesis is true, then the plots are interchangeable; however, the records from the same plot should be kept together. Technically speaking, the records done in different years in the same plot are subplots (*within-plots*) of the same main plot (*whole-plot*) and the main plots are permuted. To do this, the following options should be used during the project setup in Canoco for Windows:

Perform test: people usually use *Both above tests*. However, the procedure when one performs both tests and then uses the one that gives a better result is not correct. It may be expected that the test of the first ordination axis is stronger when there is a single dominant gradient; the test of all constrained axes together is stronger when there are several independent gradients in the data. Note that with a single explanatory variable, the first axis and all the axes are

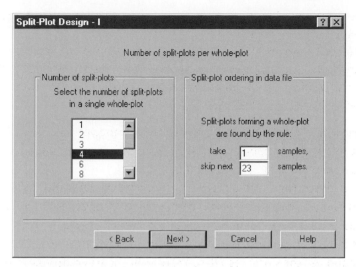

Figure 15-1. Project setup page *Split-Plot Design I.*

the same. Use of the permutation *under reduced model* is recommended. Depending on your time available, computer speed and size of the problem, the number of permutations can be increased. The permutations should be *Restricted for spatial or temporal structure or split-plot design*, then in the next setup wizard page (see Figure 15-1) named *Split-plot design*, the *Number of split-plots in whole-plot* is 4 (i.e. four records from each plot) and the split-plots are selected by the rule *Take 1 Skip 23*. This corresponds to the order of records in the file: in our case, the records from the same plot are separated by 23 records from the other plots.[*] The whole-plots are freely exchangeable and 'no permutation' is recommended at the split-plot level.[†]

After running the analysis, it is advisable to check in the Log View the permutation scheme. In your case, the output is:

```
*** Sample arrangement in the permutation test ***
   Whole plot     1 :
       1     25    49    73
   Whole plot     2 :
       2     26    50    74
   Whole plot     3 :
       3     27    51    75
etc...
```

---

[*] Take care: if you use permutation within blocks – which is **not** your case here, then the number of plots to skip is the number of plots to skip **within a block.**

[†] According to the CANOCO manual: *Optionally, the time point could be permuted also, but this . . . has a less secure basis than the permutation of sites.*

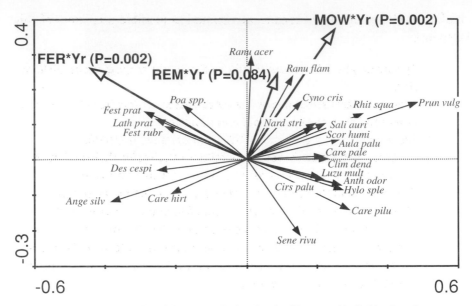

Figure 15-2. Results of the C2 analysis. The significance of individual explanatory variables is taken from the analyses C3–C5.

This shows that the whole-plots are composed of the records from the same plot, which is correct.

The relationship of particular species to experimental manipulations can be visualized by ordination diagrams. Probably the best possibility is to display the results of the analysis of *C2* by a biplot with environmental variables and species (Figure 15-2). Because you used year as a continuous variable, the interactions of time with the treatments are also continuous variables and are shown by the arrows. Because time was used as a covariable, the trends should be seen as relative to the average trend in the community. For example, the concordant direction of a species arrow with the *FER*YEAR* arrow means either that the species cover increases in the fertilized plots or that it decreases in the nonfertilized plots (or both).

15.7.   Further use of ordination results

The results of the analyses can be further used for other purposes. One of the possibilities is to use the species scores. Species scores on the constrained axis from the analyses where time*treatment was the only explanatory variable and the other factors were covariables (*C3* to *C5*) can be considered characteristics of the species response to the particular treatment. Then the following biological characteristics of species were tested as possible predictors of this response:

1. Species height, taken as the middle of the height range given in the local flora.
2. Presence of arbuscular mycorrhiza (AM), based on data from Grime et al. (1988) and from the Ecological Flora Database.
3. Seedling relative growth rate (RGR) from Grime et al. (1988).
4. Because we expected that species similar to *Molinia* would benefit most from *Molinia* removal, we also used a fourth (ordinal) variable, dissimilarity of the species from *Molinia*, for predicting the effects of *Molinia* removal. A value of 1 was assigned to graminoids higher than 50 cm, 2 to broad-leaved graminoids smaller than 50 cm, 3 to narrow-leaved graminoids smaller than 50 cm, and 4 to forbs. Spearman correlation was used for the analysis of the relationship of this value with the RDA score of *Molinia* removal.

You can easily transfer the results of analyses from the solution (.sol) file into a spreadsheet (use Excel and read the .sol file as delimited text format). Then you can use the species scores as a response and add the biological characteristics of the species you want to study (plant height, mycorrhizal status, etc.) and use them as predictor variables. It is reasonable to omit the species that have a low frequency in the data (those species show no response in a majority of sampling units, simply because they were not present before or after the treatment). Often, you will also be forced to omit the species for which biological characteristics are not available. Then you can use the data in any ordinary statistical package – we have done this using the Statistica program.

Alternatively, you can import the external variables, describing biological characteristics of species, into CanoDraw for Windows. To do so, you can use the *Project > Import variables > From Clipboard* command. When importing species characteristics, make sure you specify that the imported data refer to species, not to samples. Also, if you do not have the characteristic values available for all your species, you must import the actual data values together with species indices, identical to those used in the CANOCO project.

In our case study, plant height appears to be a good predictor of the species response to fertilization: with increased availability of nutrients, the competition for light became more important, thus the potentially taller plants increased their cover. This might be illustrated by the dependence of the species score on the constrained axis of the analysis *C3* on plant height (Figure 15-3).

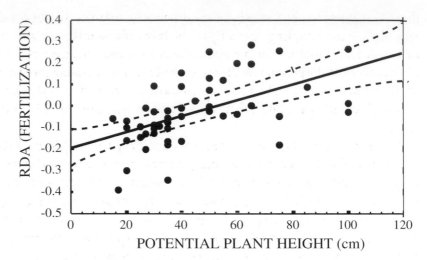

Figure 15-3. Dependence of the response to fertilization (expressed as the RDA score on the constrained axis of analysis C3) on potential plant height. The species with low frequency were omitted.

In this case, we considered the species to be independent entities and no phylogenetic corrections (Silvertown & Dodd 1996) were applied.

On the contrary, the plant mycorrhizal status has nearly no predictive power – the response of plants to fertilization does not depend on whether they are mycorrhizal or not. In a similar way, the response of plant species to the other factors can be compared with the species' biological characteristics.

15.8. Principal response curves

The method of principal response curves (PRC, see Section 9.3) provides an alternative presentation of the data analysed in this case study. The resulting response curve shows us the extent and directions of the development of grassland vegetation under different experimental treatments, compared with the control treatment. Additionally, we can interpret the directions of such composition changes using the response of individual plant species, which can be well integrated with a PRC diagram.

The vertical scores of PRC curves are based on the scores of environmental variables from a redundancy analysis (RDA), where the sampling time indicators are used as covariables and the interactions between the treatment levels and sampling times stand as environmental variables (see Section 9.3).

We will start with data stored in the Excel file *ohrazprc.xls*, in which the necessary changes have already been applied. You can compare its contents with the data used in the preceding analyses (present in the file *ohraz.xls*). Note that the main change has happened in the explanatory variables (the *Design* sheet). You will use the sampling year only in the form of a factorial variable (coded as four dummy variables *Yr0* to *Yr3*) and the experimental treatments are coded differently from the previous analyses. Each of the eight possible combinations of the three experimental treatments (mowing, fertilization and removal of *Molinia*) is coded as a separate factor level (separate dummy variable, from 0 to *MFR*).

Export the two tables (from the *Species* and *Design* sheets) into CANOCO-format data files named, for example, *prc_spe.dta* and *prc_env.dta*. Create a new project in Canoco for Windows: it will be a constrained (direct gradient) partial analysis, i.e. with both environmental variables and covariables. Specify the *prc_spe.dta* file as the species data and use the *prc_env.dta* for both the environmental and the covariable data. We suggest you name the solution file *ohrazprc.sol* and the whole project (when you are asked to save it at the end of its setup) *ohrazprc.con*. Ask for the redundancy analysis (RDA) and on the *Scaling* page select the *Inter-sample distances* and *Do not post-transform* options. Ask for log transformation of the species data, and on the next page specify centring by species; for samples, the *None* option should be selected. On the *Data Editing Choices* page, remember to check the *MAKE SUPPLEMENTARY* checkbox in the *Species* row. The *DELETE* option for *Env. variables* and *Covariables* should be pre-checked, because you specified identical file for both types of explanatory variables. Additionally, you must check the *DEFINE INTERACTIONS* option in the *Env. variables* row.

In the *Supplementary Species* page move the first species (*molicaer*) to the right list, as the species presence was experimentally manipulated so it does not represent a response of the community. In the *Delete Environmental Variables* page, move **everything** to the right list, as you will not use any of the original variables in the analysis. Instead, interactions between the treatments and years (which you will define soon) will be used. Keep only the year indicators (*Yr0* to *Yr3*) in the *Delete Covariables* page and move the other variables to the right list.

In the next setup page, you must define interactions between the sampling year indicators and the treatments. Note that the control treatment (with label 0) should be excluded and not used for treatment definitions. Therefore, $7 \times 4 = 28$ interaction terms should be defined. To create an interaction term between, say, the first year and the mowing-only treatment, select the *Yr0* variable in the *First variable* list, the M variable in the *Second variable* list, and then click the *Add* button. Continue similarly with the other 27 possible combinations. Finally, the page contents should look as illustrated in Figure 15-4. Note that the decision, which of the treatment levels should be regarded as control, is not unequivocal. Our decision to use the abandoned grassland (with no

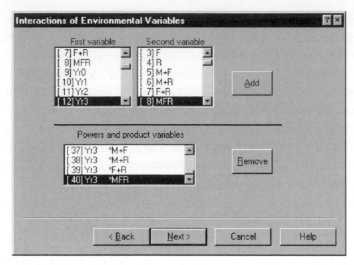

Figure 15-4. Defining interactions of time and treatment for PRC analysis.

mowing present) as a reference line is questionable and we will return to this point later.

In the *Forward Selection* page keep the *Do not use...* option, but ask for a permutation test of *Significance of first canonical axis* in the next setup page (*Global Permutation Test*). We will be testing the differences between the whole time series (four years of measurement) for each of the permanent plots. Keep the *Number of permutations* default value of 499. In the next page, select the *Restricted for...* option, and then select the *Split-plot design* option in the *Permutation Restrictions* area. The four yearly measurements for each permanent plot in the data file are interleaved with the measurements on the other plots. Therefore, in the *Split-Plot Design I* page, you must not only specify that there are four split-plots (i.e. four years of measurement) within each whole plot, but also – on the right side of the page – that to collect the four split-plots representing one whole-plot CANOCO must take 1 and skip the following 23 samples. On the next page, keep the default settings. They imply that you will randomly permute the whole plots (the permanent plots with four years of measurement) and the split-plots will not be permuted at all (the temporal ordering of samples within each permanent plot will not be changed). Use the *Finish* button to close the setup wizard and save the project under the name *ohrazprc.con*.

After you have analysed the project, switch to CANOCO Log View. At its end, you can see the results of the test for the first RDA axis, which is – in our case – also the significance test of the first principal curve. There are other items of interest in the analysis log if you calculate the PRC scores. Move up to the table of means, standard deviations and variance inflation factors for the environmental variables. You can see that all the interaction terms have identical values there, thanks to the strictly balanced experimental design. Such uniformity does not occur in every analysis; for example, if you have a different number of replicates for individual treatment levels. The standard deviation values for each interaction term are needed to calculate the PRC scores, together with the species data standard deviation, presented under the name *TAU* further up in the Log View (with value 0.59246). You can save yourself the calculations with the help of the CanoDraw for Windows program. Nevertheless, you must save the analysis log in a text file. To do so, select the *File > Save log* command from the Canoco for Windows menu. CANOCO offers the name identical to the project name, but with the *.log* extension and we suggest you accept it.

A question naturally arises as to whether just the principal response curves for the first RDA axis are sufficient to represent the trends in the community dynamics for vegetation under different experimental treatments. To answer the question, you must test the significance of the second and (if second was found significant) higher constrained RDA axes, using the method suggested in Section 9.1. Note that in this project, you already have covariables (the *Yrx* dummy variables) so you must create a new data file with covariables, containing not only the four year indicators, but also the *SamE* scores from the current analysis, exported from the *ohrazprc.sol* file.

We will not demonstrate the individual steps for testing the significance of the higher RDA axes, but the results of the tests indicate that the PRCs corresponding to the second RDA axis are not significant ($P = 0.338$ with 499 permutations).

Now, you will create a diagram with the principal response curves. Create a new CanoDraw project from the *ohrazprc.con* project (click, for example, on the *CanoDraw* button in CANOCO Project View). We will start the creation of the PRC diagram by defining the PRC scores for the first RDA axis. Select the *Project > Import variables > Setup PRC scores* command. Select all the interaction terms in the listbox, keep the default setting of importing PRC scores just for the first RDA axis, and locate the *ohrazprc.log* file using the *Browse* button. The final look of the dialog contents should resemble the contents of Figure 15-5.

After you click the *OK* button, the PRC scores are created and stored among the imported variables. Note, however, that you must still set up the horizontal coordinates for the PRC diagram. These correspond to a quantitative time (year) value for each combination of treatment and year. It is important to realize that the individual points in this diagram **do not** correspond to project samples, but rather to individual environmental variables (their interaction

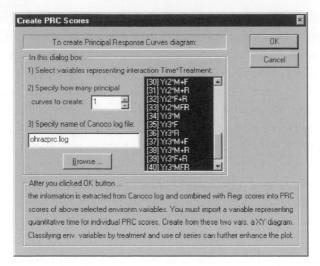

Figure 15-5. Importing the PRC scores into a CanoDraw project.

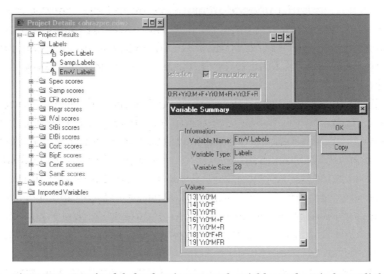

Figure 15-6. Copying labels of environmental variables to the Windows Clipboard.

terms). The easiest way of creating a variable with the required X coordinates is described in the following paragraphs.

Select the *View > Project Details* command and expand the *Project Results > Labels* folder in the new window which appears after that. When you click the item labelled *EnvV.Labels* with the right mouse button, CanoDraw displays a summary of this particular variable (which stores the labels for environmental variables). Click the *Copy* button in the *Variable Summary* dialog to copy both indices and labels of environmental variables to the Windows Clipboard (see Figure 15-6).

Open a new document in Microsoft Excel and paste the data from the Clipboard. Two columns of data are pasted. The first one contains the indices of individual environmental variables (they do not start at 1, but with 13, as the 12 original environmental variables were deleted in the CANOCO project setup), the second column contains the names of the interaction terms. You will use the second column to guide you about the appropriate value of the *Time* variable, which you will now create in the third column. Write the new variable name ('Time') at the top of the third column and write 0 in all rows that involve *Yr0* in the interaction term, and similarly 1 for terms with *Yr1*, etc. The exact ordering of terms depends on the order in which you defined them in the CANOCO project setup wizard.

After you have completed the third column (with seven 0s, seven 1s, seven 2s and seven 3s), delete the middle column, because labels cannot be imported back into a CanoDraw project. Select the two remaining columns (only the first 29 rows with column labels and data) and copy them to the Windows Clipboard. Switch back to the CanoDraw program and select the *Project > Import variables > From Clipboard* command. CanoDraw displays a dialog box in which you must change the option on the left side from *Samples* to *Environm. variables*. Keep the option *First column contains indices* selected (checked). Then you should click the *Import* button.

Additionally, you need to classify the environmental variables into seven classes, depending on which treatment they belong to. Use the *Project > Classify > Env. variables* and click the *New select* button. In the *Manual Classification* dialog, you must define the seven classes corresponding to individual treatments (use the *Add class* button) and place the interaction terms in the appropriate classes (keep the *Continue with Class Members dialog* option checked when defining the class name), as illustrated in Figure 15-7. When you close the *Manual Classification* dialog, make sure you selected the *Use this classification in diagrams* option.

Finally, to create real curves, not only a scatter of points, you must define a series collection for environmental variables. Each of the treatment types will become a separate series in the collection, with the four interaction terms arranged in the order of increasing year value. You can use the just-created classification to make the definition of series collection easier. Select the *Project > Define Series of > Env. variables* command and select the *From class* button. Select the (only one) classification offered in the next dialog. Check in the *Series Collection* dialog box that the four interaction terms within each series are ordered so that the year value increases. If not, drag the items to their appropriate position in the right-hand listbox. After you close the dialog with the *OK* button, make sure that the *This collection is used . . .* option is checked.

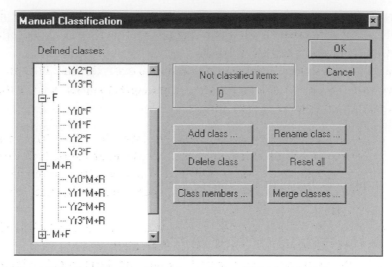

Figure 15-7. Classification of environmental variables based on the treatment type.

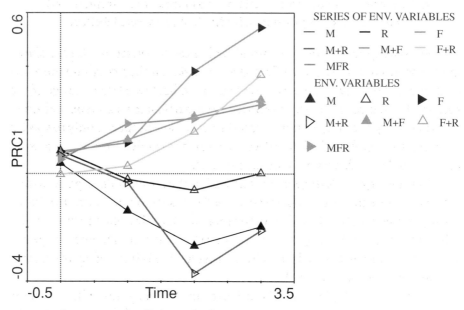

Figure 15-8. PRC1 diagram, the first attempt.

To create the PRC diagram, select the *Create > Attribute Plots > XY(Z) Plot* command. Select the *Imported data > For env. variables > Time* variable in the *X VARIABLE* field and, similarly, the *PRC1* variable in the *Y VARIABLE* field. Select the *None* option in the *PROPERTIES > Labelling* area and also *None* in the *VISUALIZATION MODEL* area. After you click the *OK* button, the diagram is created (see Figure 15-8). If you do not see the legend at the bottom of your diagram, you can enable it in the *Appearance* page of the *Project* > *Settings* dialog box. Note however, that the diagram is somewhat inefficient in reproducing the results, so we will modify it anyway.

We will not describe all the changes needed to make the final PRC diagram, but we will outline them here (the resulting diagram is part of the extended PRC diagram, shown in Figure 15-10):

1. We encoded the treatments by line type, not by line colour. The lines for treatments where fertilization was involved are substantially thicker than the other lines. The lines for treatments including the removal of *Molinia* have a dashed style, and lines where the mowing treatment was included are grey, not black.
2. We have added the reference line for control treatment to the legend and also to the plot (it overlays the horizontal axis line).
3. Symbols were removed from both the plot and the legend. The legend frame was removed and the layout made more condensed.
4. The labelling of the horizontal line was deleted (the minimum and maximum) and the positions of individual years were labelled.

The PRC diagram can be substantially enhanced by a vertical 1-D plot, show-ing the scores of species on the first RDA axis. We can then combine the value read from a PRC curve with a species score to predict the relative change of the species abundance (or cover, in our case) at a particular treatment and time. Before creating the plot, we must limit the set of plotted species only to species with a reasonable fit to the first RDA axis. It can be assumed that we can predict the expected abundances only for such species.

Select the *Project* > *Settings* command and in the *Inclusion Rules* page change the *Species* value in the *Lower Axis Minimum Fit* area from 0 to 7. Only 20 'best-fitting' species will be selected from the total of 86 species. Note that use of the values in the *Species Fit Range* area is not correct in this context, because species with their abundances fitted badly by the first RDA axis but well by the second RDA axis may become selected.

To create a 1-D plot, we should use an *XY* diagram with constant *X* coordinates. Note, however, that CanoDraw will, by default, stretch even the negligible range of the *X* values in such a diagram. We need to control the shape of the resulting diagram, so that it is tall and narrow. To do so, select first the *View* > *Diagram Settings* command and check the *Adjust graph aspect ratio* in the *Properties 1* page. While here, we can make the vertical axis of the upcoming diagram more informative by switching to the *Properties 2* page, checking the option *All tickmarks with labels*, and un-checking (deselecting) the option *Labels on vertical axis are rotated*.

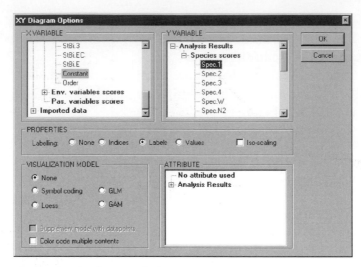

Figure 15-9. Creating a vertical 1-D plot with plotted species scores on the first RDA axis.

Now select the *Create > Attribute Plots > XY(Z) Plot* command. In the *X VARI-ABLE* area, select the *Analysis Results > Species scores* and there the *Constant* item near the bottom of the list. In the *Y VARIABLE* area select *Analysis Results > Species scores* and then *Spec.1* item. Keep the other options at their default values (see Figure 15-9). When you click the *OK* button, CanoDraw first displays a dialog named *Aspect ratio for 45-deg. banking*. Change the default value to 8. This will make the diagram a rectangle eight times higher than is its width.

We polished the resulting diagram to some extent* before combining it with the modified PRC diagram. The final presentation of the PRC analysis is displayed in Figure 15-10.

The PRC diagram shows that there are two directions of departure from the vegetation composition on the reference plots (which are not mown, no fertilizer is added to them and *Molinia caerulea* is not removed from them). The mown plots (with negative PRC scores) have a higher abundance of *Nardus stricta*, of several small sedges (*Carex pallescens*, *Carex panicea*), and also of many forbs. Also the mosses (such as *Aulacomnium palustre* or *Rhytidiadelphus squarrosus*) have a much higher cover. The extent of the mowing effect is somewhat smaller than the oppositely oriented effect of fertilization, shown by the lines directed to the positive side of the vertical axis. The fertilized plots become dominated by

---

* Removed *Spec.1* label for vertical axis; removed horizontal axis labels; removed right-hand vertical axis line; changed species symbols, and adjusted labels position to the right side of the reference line. The two diagrams (with curves and species) were finally combined in the Adobe Photoshop program. Note that only the zero values on the vertical axes must be aligned, but not the scale of the axes.

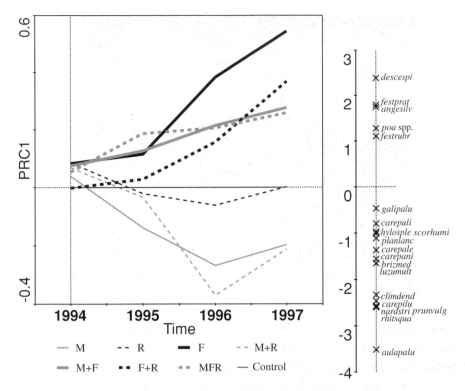

Figure 15-10. Final diagram with principal response curves.

competitive grass species (*Deschampsia cespitosa, Festuca pratensis, Poa* spp.) with only a few forbs (e.g. the tall forb *Angelica silvestris*) increasing their cover. There is only a limited effect of dominant grass removal, and it is similar to the effect of mowing (the dashed lines are generally at lower positions than the corresponding solid lines).

The extent of increase in the cover of grasses such as *Festuca pratensis, Poa* spp. or *Festuca rubra* can be quantified using the rules specified in Section 8.3.11.2 of the CANOCO reference manual (Ter Braak & Šmilauer 2002). The scores of the three grass species are around the value of $+1.5$. If we look up the PRC score of the fertilized-only plots (F in the diagram) in the year 1997, we see that it is approximately $+0.5$. The estimated change is, therefore, $\exp(1.5 \cdot 0.5) = 2.12$, so all the three grasses are predicted to have, on average, more than two times higher cover in the fertilized plots compared with the control plots.

The variability of PRC scores in the year 1994, observed on the left side of the diagram, cannot be caused by the experimental treatments (which were started afterwards), so it provides a 'yardstick' to measure the background variability among the plots. This suggests that there is probably no real difference

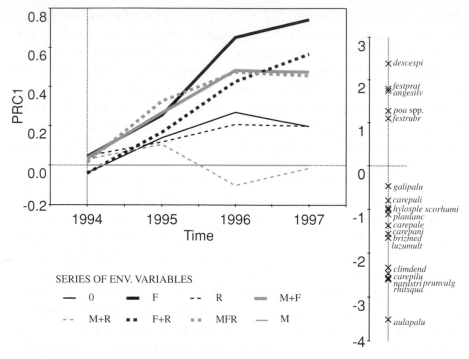

Figure 15-11. PRC diagram for the first RDA axis. The mowing-only regime was taken as the reference (control) treatment.

among the three treatments, where fertilization was combined with mowing and/or with removal of *Molinia*. Likewise, the mown plots did not differ from those where the removal treatment was added to mowing.

We have already discussed at the beginning of this section how the unmown plots may not be the best choice for control treatment. If you want to use the **mown** but unfertilized plots without *Molinia* removal as the reference level, the only change to be made in the CANOCO project is to remove the interaction terms including the M treatment, and add instead four new interactions between year indicators and the 0 treatment.

If you create a new PRC diagram using the modified analysis (see Figure 15-11), the only change is that the identity of the reference line, which is 'rectified' (flattened) along the horizontal axis, has changed. The scores of species on the first RDA axis (the right side of the plot) have not changed at all and this assures us that the interpretation of the PRC curves has not changed either.

Finally, we can compare the information you obtain from summarizing the standard constrained ordination (as presented in Section 15.6) using an ordination diagram such as the biplot diagram in Figure 15-2, and that from the

PRC diagram (e.g. Figure 15-10). It should be noted that the analyses and corresponding diagrams are not the only possible ones and you should probably run more of them when evaluating your research data. For example, the constrained analyses with time coded as a categorical variable could also be graphed, and the partial PRC analysis can be constructed for each of the factors separately, using the other ones as covariables. Also, you could compare the species scores on the first principal response axis with species traits (and you would find that species with positive scores, i.e. those most positively affected by fertilization, are the tall species).

However, when submitting a paper manuscript, one usually has very restricted space, so only a small selection of the performed analyses can be presented in graphical form: only the most instructive diagrams are needed. The PRC diagram is superior in its display of temporal trends. In our example, the PRC diagram clearly shows that the development of plots diverges according to the treatment during all the years, particularly for the fertilized plots (and in this way confirms that the use of time as a quantitative variable is a good approximation). In a longer run of the experiment, we would probably be able to find some stabilization and could estimate the time needed to achieve a 'stable state'. On the contrary, the classical diagram is better at showing the affinities of individual species to the treatments (which one is more responsive to mowing, and which to fertilization), and also the mutual relationships of the effects (the similarity of effects of mowing and dominant removal). The information presented in the two diagrams is partially overlapping, and partially complementary.

Based on the combined information from all the analyses, we would conclude that many species (particularly the small ones) are suppressed by either the fertilization or by the absence of mowing or by a combination of both factors. Only a limited number of (mostly tall) species is favoured by fertilization.

Case study 6: Hierarchical analysis of crayfish community variation

In this chapter, we will study the hierarchical components of variation of the crayfish community in the drainage of Spring River, north central Arkansas and south central Missouri, USA. The data were collected by Dr Camille Flinders (Flinders & Magoulick 2002, unpublished results). The statistical approach used in this study is described in Section 9.2.

16.1. Data and design

The species data consist of 567 samples of the crayfish community composition. There are 10 'species', which actually represent only five crayfish species, with each species divided into two size categories, depending on carapace length (above or below 15 mm). Note that the data matrix is quite sparse. In the 5670 data cells, there are only 834 non-zero values. Therefore, 85% of the data cells are empty! This would suggest high beta diversity and, consequently, use of a unimodal ordination method, such as CCA. There is a problem with that, however. There are 133 samples without any crayfish specimen present and such empty samples cannot be compared with the others using the chi-square distance, which is implied by unimodal ordination methods. You must, therefore, use a linear ordination method.

The sampling used for collecting data has a perfectly regular (balanced) design. The data were collected from seven different watersheds (WS). In each, three different streams (ST) were selected, and within each stream, three reaches (RE) were sampled. Each reach is (within these data) represented by three different runs (RU) and, finally, each run is represented by three different samples. This leads to the total of 567 samples = 7 WS · 3 ST · 3 RE · 3 RU · 3 replicates. These acronyms will be used, when needed, throughout this chapter.

Table 16-1. *Analyses needed to partition the total variance in the crayfish community data*

Variance component	Environmental variables	Covariables	Permuting in blocks	Whole-plots represent
Watersheds	WS	None	No	ST
Streams	ST	WS	WS	RE
Reaches	RE	ST	ST	RU
Runs	RU	RE	RE	None
Residual	None (PCA)	RU	n.a.	n.a.
Total	None (PCA)	None	n.a.	n.a.

Both data tables are stored in the *scale.xls* Excel file, species composition data in the *species* sheet, and the variable describing the sampling design in the *design* sheet.

When performing the variance decomposition, you must partition the total variance into five different sources, using the method summarized in Table 16-1.

Note that in the permutations tests assessing the effects of spatial scales (watersheds, streams and reaches) upon the crayfish community, the individual samples cannot be permuted at random. Rather, the groups representing the individual cases of the spatial levels immediately below the tested level should be held together. This can be achieved using the split-plot design permutation options.

16.2. Differences among sampling locations

You will start your analyses with a 'standard' PCA, in which you will not constrain the ordination axes and you will not use any covariables. Instead, the indicator variables for the watersheds and streams are passively projected to the ordination space to visualize the degree of separations at the two highest spatial levels. Note that you could also use the indicators (dummy variables) for the lower spatial levels, but the resulting diagram would be too overcrowded.

The first step you must perform is to export the data sets into CANOCO data files, using WCanoImp. You can use the names *scale.sp.dta* and *scale.en.env* for the data files. Note that in the *design* sheet, the first six columns (*date* to *sample*) do not have an appropriate coding, so you will not use them in your analyses.

Create a new CANOCO project named *cray.tot.con* and specify an indirect (unconstrained) analysis with *scale.en.dta* used as environmental variables, which are passively projected to the resulting ordination space. Analysis type is PCA and you should keep the default values for the other options. Ask for deletion of environmental variables and in the *Delete Environmental Variables* wizard page keep only the variables 9 to 36 (*English* to *West3*) in the left-hand column. Consequently, only the positions of the seven watersheds and the 21 (7·3) streams will be projected to PCA ordination space.

Create a new project for this analysis in the CanoDraw program and start by specifying all the retained environmental variables as nominal (*Project* > *Nominal variables* > *Environmental variables*). Some of the crayfish species are not well correlated with the main gradients of species composition (principal components), so their presence in the diagram provides limited information. To suppress them, select the *Project* > *Settings* command and in the *Inclusion Rules* page change the 0 value in the *From* field of the *Species Fit Range* area to 2. This change excludes three of the 10 'species', that have less than 2% of their variability explained by the first two PCA axes.

Create a biplot with species and environmental variables (they are presented as centroids of samples, because you specified them as nominal variables) using the *Create* > *Biplots and Joint Plots* > *Species and env. variables* command. Note that the diagram content is dominated by the difference of the *Bay* watershed from the other ones, so most of the location centroids are concentrated in the middle of the diagram. Therefore, we present here the same information split into two diagrams. In the first (Figure 16-1) diagram, only the centroids for watersheds are shown, together with the arrows for selected species. This diagram indicates that the main gradient of species composition change is related to increasing abundance of the *punct* crayfish species, and that the difference of the samples from the *Bay* watershed can be explained by the more frequent occurrence of the two '*oz*-X' taxa.

The other diagram (Figure 16-2) shows only the central part of the previous ordination diagram, and shows centroids not only for the watersheds (black squares), but also for each of the three streams within each watershed (grey squares), except the *Bay, Bay2* and *Wf3* centroids, cut off from the diagram.

You can see that while the positions of some stream triples are clustered in the proximity of their parental watershed, the difference among the streams is much higher for others (such as *Bay*, but also *Pinehill*).

The relevant part of the analysis log for this PCA is shown here:

```
*** Summary ****

Axes                                    1       2       3       4    Total variance

Eigenvalues                         : 0.382   0.192   0.128   0.096         1.000
Species-environment correlations :  0.472   0.745   0.640   0.599
Cumulative percentage variance
   of species data                 : 38.2    57.4    70.2    79.8
   of species-environment relation: 27.3    61.6    78.5    89.5

Sum of all         eigenvalues                                              1.000
```

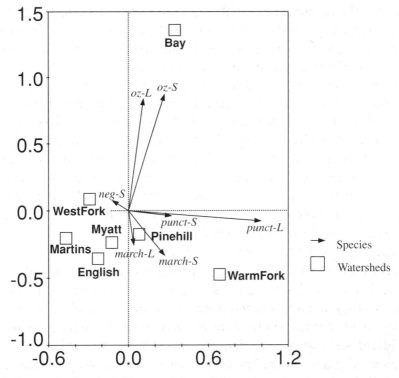

Figure 16-1. First two PCA axes, with crayfish species and projected watershed centroids.

The total variation in the species data is, of course, equal to 1.0, because CANOCO standardizes in this way the species data values in linear ordination methods. This will make it easier for us to read the explained variance fractions in the other analyses. The first two principal components explain more than 57% of the total variation in crayfish community composition.

16.3.   Hierarchical decomposition of community variation

Now you need to create a separate CANOCO project for each of the spatial levels you want to evaluate (see Table 16-1). Note that, in fact, you can omit calculations for one of the levels, because the fractions sum up to 1.000. Probably the best choice for omission is the residual variation (within-runs variance), as there is no meaningful permutation test and defining the indicator variables (to be used as covariables in the partial PCA) for individual runs is quite tedious.* We will not provide detailed instructions for all the projects

---

* You will need to do exactly this anyway in the analysis of the effect at runs level, where the runs will represent environmental variables.

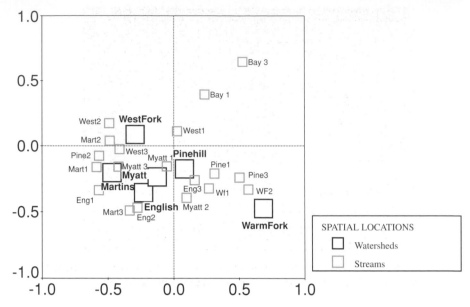

Figure 16-2. First two PCA axes, with projected centroids of watersheds and streams.

you must set up here, but if you feel confused you can always look at the corresponding CANOCO projects, which are defined in the sample data collection (see Appendix A).

Start with the analysis described in the first row of Table 16-1. This analysis has no covariables, and the seven watershed indicator variables are used as the environmental variables (see Figure 16-3).

Do not ask for forward selection of environmental variables, but in the *Global Permutation Test* page ask for *Significance of canonical axes together*. We cannot *a priori* expect for only the first canonical axis to summarize all the differences in crayfish community composition among the watersheds (or the other kinds of location in the following analyses), so we should base our test on the whole canonical space. We suggest you specify only 199 permutations, as the calculations for this large dataset can take a relatively long time. On the next setup page, ask for *Restricted for spatial or temporal structure or split-plot design* and select the *Split-plot design* option on the page after that.

When you arrive at the *Split-Plot Design I* setup wizard page, you must specify the size of whole-plots. As you remember from the introductory section, we will permute individual streams, keeping all the reaches of a particular stream together. Therefore, all the samples within each stream represent one whole-plot. Therefore, there are 27 split-plots within each whole-plot (3 repl · 3 RU · 3 RE), so select this number. The samples representing one

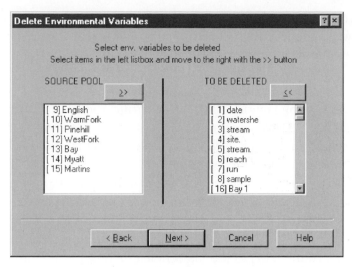

Figure 16-3. The watershed indicator variables are retained as environmental variables when testing variation explained at the watershed level.

whole-plot are contiguous in the dataset, so you can keep the default settings (*take* 1, *skip next* 0) on the right side of the setup wizard page (using *take* 27, *skip next* 0 would have the same effect, of course).

In the next, *Split-Plot Design II* page, specify free exchangeability of the whole-plots (streams) and no permutation for the split-plots. After you analyse the project (*crayf.ws.con*), you can determine the amount of variability explained by the streams from the Summary table, using the *Sum of all canonical eigenvalues* row (the last one).

```
**** Summary ****

Axes                                    1      2      3      4    Total variance

Eigenvalues                        : 0.075  0.059  0.039  0.011        1.000
Species-environment correlations   : 0.540  0.565  0.427  0.329
Cumulative percentage variance
   of species data                 :   7.5   13.4   17.3   18.5
   of species-environment relation:  39.7   71.0   91.7   97.8

Sum of all           eigenvalues                                      1.000
Sum of all canonical eigenvalues                                      0.189
```

The watersheds explain 18.9% of the total variability in crayfish community data. CANOCO also reports in the analysis log about the permutation test results. Because you specified some restriction on the permutations (the split-plot design), CANOCO displays the indices of samples that were assigned to individual parts of the design structure. This is useful for cross-checking the correctness of the specified permutation structure. The actual permutation test results show that none of the 199 permuted (randomized) datasets produced as large an amount of variability explained by watersheds as the true data configuration, because we achieved the smallest possible type I error estimate (i.e. $(0 + 1)/(199 + 1) = 0.005$).

The next step (second row in Table 16-1) quantifies the variability explained at the streams level. We will use the ST variables (*Bay1* up to *West3*) as the environmental variables and the seven WS variables (the next higher spatial level) as the covariables. Because we do not want to involve the differences between watersheds in the permutations simulating the null model, we must permute the whole-plots (the next lower spatial level, i.e. the reaches) randomly only **within** each of the watersheds. Therefore, we will not only have the split-plot design restrictions on the permutations, but we will also permute in blocks defined by the covariables. So the important difference in the new project (*crayfst.con*) setup is (in addition to specifying different environmental variables and adding the watershed indicators as covariables) in the *Permutation Type* page, where you should check the *Blocks defined by covariables* option (and select the *Restricted for . . .* option, of course). A new setup page will appear, asking you to select which covariables define the blocks. You should select all seven variables here. In the *Split-Plot Design I* page, the number of split-plots in each whole-plot is 9 (3 repl · 3 RU, in every reach).

You can see in the analysis results that streams explain 12.3% of the total variability and that this is a significantly non-random part of the total variation ($P = 0.015$).

The next analysis is similar to the one you just performed, except you move one level down in the hierarchical spatial structure: the placement of samples in reaches (coded by variables with their names ending with R1, R2 and R3) is used as environmental variables, and the 21 stream indicators are used as covariables. You should permute within blocks defined by these covariables and use the split-plot structure, where the whole-plots are the individual runs (each with three samples — split-plots). The variability explained at the reaches level is estimated as 15.8% of the total variability and the type I error of the permutation test is estimated as $P = 0.005$.

The last variance component you will estimate is the variation between the runs. The runs are not coded with individual dummy variables in the *scale.en.env* data file, so you must delete **all** the environmental variables in the setup wizard, and ask for definition of interaction terms for environmental variables. In the *Interactions of Environmental Variables* page, you should define all possible pair-wise interactions between the 63 reaches and the three variables named *Run1*, *Run2* and *Run3*. Note that the *Runx* variables code for run membership

Table 16-2. *Results of variance decomposition*

Component	Explained variability (%)	DF	Mean square value	Test significance
WS	18.9	6	3.150	0.005
ST	12.3	14	0.879	0.015
RE	15.8	42	0.376	0.005
RU	19.5	126	0.155	0.005
Residual	33.5	378	0.089	n. a.
Total	100.0	566	0.177	n. a.

within the context of each individual reach (i.e. the first run within a reach has values 1, 0, 0 for these three variables for every reach). Therefore, you must define 189 interaction terms here! When specifying the permutation you should ask for unrestricted permutations (i.e. no split-plot design), but **within** the block defined by covariables (i.e. within the individual reaches). The variation explained at the runs level is 19.5% of the total variance. The significance level estimate is $P = 0.005$.

Finally, you can deduce the amount of variability explained by the differences between the samples within runs by a simple calculation: $1.0 - 0.189 - 0.123 - 0.158 - 0.195 = 0.335$.

The results of all the analyses are summarized in Table 16-2.

The table includes not only the absolute fraction of the explained variation, but also the values adjusted by the appropriate number of degrees of freedom. These are calculated (similar to a nested design ANOVA) by multiplying the number of replications in each of the levels above the considered one by the number of replications at the particular level, decreased by one. Note that by the number of replications, we mean number **within** each replication of the next higher level. For example, there are seven watersheds, each with three streams, and each stream with three reaches. Therefore, number of DFs for the reach level is $7 \cdot 3 \cdot (3 - 1) = 42$.

It is, of course, questionable which of the two measures (the variation adjusted or not adjusted by degrees of freedom) provides more appropriate information. On one hand, the reason for adjusting seems to follow naturally from the way the sum of squares is defined.* On the other hand, it seems quite natural to expect the increasing variation at the lower hierarchical levels: there

---

* It does not matter here that we use the data which were standardized to total sum of squares equal to 1. All the components are calculated with the same standardization coefficient – the total sum of squares of the data does not vary across the individual projects.

must always be more reaches than there are streams or watersheds, so a higher opportunity for crayfish assemblages to differ.

In any case, the variation at the watershed level seems to be relatively high compared with the other spatial scales, and all the spatial scales seem to say something important about the distribution of crayfishes, given the fact their 'mean-square' terms are substantially higher than the error mean-square (0.089).

17

Case study 7: Differentiating two species and their hybrids with discriminant analysis

While we can think of potential applications for linear discriminant analysis (LDA) in ecology (such as finding differences in habitat conditions among *a priori* recognized community types), its real use is quite infrequent. This is probably caused by the fact that many problems that seem to be a good fit for discriminant analysis produce weak rather than convincing results. A good example of this might be an attempt to use discriminant analysis for the selection of plant species indicating a particular community type. The discriminant analysis then attempts to separate the community types using a linear combination of species abundances. However, because the species occurrences are often strongly correlated and each of them alone provides little information, it is usually impossible to single out few such diagnostic species, even if the species composition differs strongly among the distinguished types. We will therefore use an example from a numerical taxonomy here, in the hope that the topic will be easily understood by ecologists and that they can apply the hints provided to their own problems.

17.1. Data

The sample data were taken from a taxonomic study[*] of several species of *Melampyrum*, hemiparasitic plants from the Scrophulariaceae family. Eighty plant specimens were measured, selected from four taxonomic groups: *M. bohemicum* from Czech localities, *M. bohemicum* from Austrian localities, hybrid populations of *M. bohemicum* × *M. nemorosum*, and *M. nemorosum*. Each group was represented by 20 specimens, originating usually from several local populations. You will be looking for the plant vegetative and

---

[*] Štech (1998); only a small subset of the collected data is used in our case study.

flower-related characteristics, which would allow the four taxonomic groups to be distinguished.

The data are contained in the *melampyr.xls* Excel spreadsheet file. Everything is contained in a single sheet, the actual measurements coming first, followed by the information important for a post-analysis modification of the biplot (*BipE*) scores of selected characteristics, which will be described later in this chapter.

The first column, named *Group*, identifies the taxonomic group to which the particular observation (row) belongs. Note that this coding cannot be used in CANOCO, so the same information is present in expanded form in the next four columns. The individual levels (1–4) of the *Group* variable are re-coded into four dummy (0/1) variables, which also have more informative names. The next 22 columns (*S1* to *SDU2*) contain various types of measurements taken mostly on the flowers or on the bracts, which are parts of the inflorescence. The measurement unit is millimetres. We will not explain the meaning of individual characteristics here, but those selected for discrimination will be characterized later in the chapter.

CANOCO requires that the response and the explanatory variables reside in separate input files. Therefore, we will export the classification of plants into one file and the values of the 22 morphological characteristics into another one. Select the data in columns B to E and copy them to the Windows Clipboard. Store the classification in a file named *mel_clas.dta*, using the WCanoImp program. Remember to check the [*Generate labels for*] *Samples* option, as the individual rows have no labels. Export the morphological measurements in spreadsheet columns *F* to *AA* to another file, named *mel_char.dta*, in a similar way. It may be worthwhile considering log-transformation of the measured dimensions, but we will ignore this issue in our tutorial.

17.2.   Stepwise selection of discriminating variables

Create a new CANOCO project – we will use species data (*mel_clas.dta*, the classification of the 'samples' into four groups) and environmental variables (*mel_char.dta*, the morphological measurements), and specify a constrained analysis on the same page. Specify the names of data files and *mel_cva.sol* as the name of the solution file (and, later on, *mel_cva.con* as the name of CANOCO project) on the next page. Specify canonical correspondence analysis (CCA) on the following page. In the *Scaling* page of the setup wizard, specify focus on *Inter-species distances* and the *Hill's scaling* option. Keep the default settings for the other options, up to the *Forward selection* page, where you should specify the *Manual selection* option. Make sure that the use of Monte Carlo

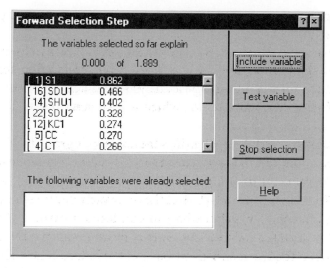

Figure 17-1. Dialog displaying first step of the forward selection procedure.

permutation tests is enabled, and specify 999 for the number of permutations. On the next wizard page, specify unrestricted permutations.

Click the *Analyze* button to start the stepwise selection of discriminating variables. When the *Forward Selection Step* dialog appears for the first time (the list with selected variables at the dialog bottom is empty), the marginal (independent) effects of individual characteristics are shown (see Figure 17-1).

You can see that the discrimination ability differs substantially among the individual characteristics. The variable *S1* (width of lowest bract) explains 0.862 from the 1.889 of the total explainable inertia (i.e. almost 46% of the total variation explainable by all the 22 characteristics). If you wish, you can check the significance of the marginal effects of individual measured characteristics by selecting in turn each characteristic in the list and clicking the *Test variable* button. Be careful not to click the *Include variable* button (or to press the *Enter* key, because that button is selected by default). When you include a variable, the inclusion is irreversible for the particular analysis run. If you check the marginal effects, you can see that almost all the variables have some discriminating power, except the last four characteristics in the list (*DDP*, *KT2*, *OT* and *V1*).

If you did not perform the tests, do the test on the first (the best) candidate variable *S1* now. A dialog named *Report on Permutation Test* appears. As you can see the type I error probability value is estimated to be 0.0010. This is, of course, the lowest achievable value when we use 999 permutations. Select this variable using the *Include variable* button. CANOCO moves the variable into the lower list (and shows it there with the recently estimated significance level). As you can see, ordering of the remaining candidate characteristics changed somewhat – those that were more correlated with the *S1* variable decreased in their importance after *S1* was selected (particularly the *SDU2* variable).

You can continue with the selection of the best candidate in each step (the one at the top of the list) until the type I error probability exceeds a threshold selected *a priori*. The question is which value to use as the threshold. If you want to have a well-discriminating group of descriptors that have a significant relationship with taxonomic group membership at the level 0.05 and you accept the reasoning behind the adjustment of acceptance thresholds at the partial decision steps (the Bonferroni correction, see Section 5.9 for additional comments), there are at most 22 decisions to be made during the selection of 22 predictors. Therefore, with the *a priori* selected threshold level of 0.05, you should accept individual predictors with estimated type I error probabilities below the $0.05/22 = 0.0023$ value. When you apply this threshold to these data, you should accept, in addition, the *SHU1* variable (width of upper calyx teeth) and the *CT* variable (length of corolla tube), both with the type I error estimates equal to the lowest achievable value (0.001). But because the primary goal of discriminant analysis is to create reliable discriminating rules, we suggest that a little redundancy won't hurt (at least for a variable with significant marginal effect), so we suggest you also accept the next candidate, the *DHP* variable (length of the upper lip of the corolla), despite its significance estimate of about 0.025. Note that these four best variables explain more than 79% (1.498 of 1.889) of the variation that would be explained by all the 22 descriptors together. End the forward selection now using the *Stop selection* button.

CANOCO displays in its Log View the Summary table of the performed analysis:

```
**** Summary ****
                                                                Total
Axes                                    1      2      3     4    inertia

Eigenvalues                          : 0.872  0.406  0.221  0.779  3.000
Species-environment correlations     : 0.934  0.637  0.470  0.000
Cumulative percentage variance
    of species data                  : 29.1   42.6   49.9   75.9
    of species-environment relation: 58.2   85.3   100.0  0.0

Sum of all           eigenvalues                                 3.000
Sum of all canonical eigenvalues                                 1.498
```

Note that to obtain the eigenvalues traditionally used to characterize the discriminant scores (θ), you must transform the values provided in this summary

(λ) using the formula:

$$\theta = \lambda/(1 - \lambda)$$

which is 6.81, 0.68 and 0.17 for the three canonical (discriminant) axes. Note that the *Total inertia* is equal to the number of groups minus one.

17.3.   Adjusting the discriminating variables

Before we can create a CanoDraw project from this canonical variate analysis (CVA) project, we should define an additional CANOCO project, which we will use to obtain the information needed to adjust the lengths of environmental variables' arrows, so that they better express their discriminating power (see Section 8.4.3.1 of Ter Braak & Šmilauer 2002, for additional discussion). The biplot scores in a standard ordination diagram enable comparison among the environmental variables using the variable standard deviation as the scaling unit. But for the discriminant analysis, a better reference scale is provided by the within-group standard deviation (i.e. the within-group variability of the environmental variables). To correct the lengths of arrows, represented by the values in the *BipE* scores section in the solution file from the project just analysed, you must obtain support information about the between-group variation of the individual predictor variables (see Section 8.4.3.1 of Ter Braak & Šmilauer 2002, for more details).

Define a new CANOCO project (named, for example, *mela_rda.con*, and producing the *mela_rda.sol* solution file). Specify constrained analysis – RDA – using the *mel_char.dta* as the species data and *mel_clas.dta* as the environmental data (note that this assignment is reversed compared with the preceding discriminant analysis). Specify post-transforming of species scores (dividing by standard deviations) and scaling focused on inter-species correlations, and centring by species. In the *Data Editing Choices* page specify that you want to *DELETE* some *Species*. Then, in the *Delete Species* page, delete all the characteristics except the four retained during the forward selection in discriminant analysis (i.e. *S1*, *CT*, *DHP* and *SHU1*). Select neither forward selection nor a permutation test. Analyse the project and look up in the resulting solution file the *CFit* section. The %EXPL column contains the information we need to calculate correction factors for the *BipE* scores in the *mel_cva* solution file. The required calculations are illustrated in Table 17-1.

You must calculate the f factor (as 100 / (100–%EXPL) for each environmental variable) and then multiply its square-root by the *BipE* score-values in the other solution file (*mel_cva.sol*) to obtain the modified *BipE* score. The

Table 17-1. *Calculating adjustment factors for biplot scores, using the information from the CFit section of the mel_rda.sol file. Columns not relevant to calculations were omitted*

N	Name	% EXPL	f	sqrt(f)
2	S1	86.2	7.246	2.692
5	CT	26.55	1.361	1.167
8	DHP	23.52	1.308	1.143
15	SHU1	40.21	1.673	1.293

Table 17-2. *Original and transformed biplot scores of species traits*

N	Name	Biplot score					
		Original			Modified		
2	S1	0.3544	0.1059	0.0343	0.9540	0.2851	0.0923
5	CT	0.0360	0.3861	0.7398	0.0420	0.4506	0.8633
8	DHP	−0.1045	0.4857	0.0013	−0.1194	0.5552	0.0015
15	SHU1	0.0924	0.7095	0.0486	0.1195	0.9174	0.0628

original and the modified *BipE* scores are shown in Table 17-2. Note that you must open the *mel_cva.sol* file in the Notepad program and **replace** the numbers present there with the new ones, using the same number of decimal digits.

Then you can save the modified solution file, close the *mel_rda.con* project, and start CanoDraw from the original, *mel_cva.con* project. CanoDraw will read the changed solution file and plot the morphological characteristics with the modified arrow lengths.

17.4.   Displaying results

To plot the first two discriminant axes, you should just use *Create > Biplots > Biplots and Joint Plots > Species and env. variables*. Before doing that, however, select the *View > Diagram Settings* command and make sure that in the *Properties 1* page the *Show rescaling coefficients . . .* option is selected. While there, you can uncheck the *Rescale sample or species scores to optimality* option, as we will need this option disabled for the second diagram we create in this project. While you create the biplot with species (the taxonomic groups) and environmental variables (the selected morphological characteristics), the *Rescaling of ordination*

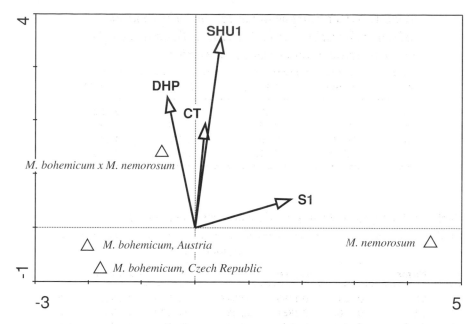

Figure 17-2. Diagram displaying first two discriminant axes. The discriminating characteristics and the centroids of classes are shown.

scores dialog appears, with a value of 1 suggested for both the *Species* and *Explanatory variables*. We suggest you change the scaling factor for explanatory variables to, say, 5, to make the arrows longer. The resulting plot (with adjusted label positions and text of the group labels) is shown in Figure 17-2.

It is obvious that the two involved species can be distinguished primarily using the *S1* characteristic, which has higher values for *M. nemorosum*. The hybrids differ from both parental species in higher values of *DHP*, *CT* and *SHU1*.

To evaluate the degree of separation among the groups of plants, we must plot the diagram with sample scores. Note that the Mahalanobis distances (Mahalanobis 1936) among the individual observations are preserved when the scaling is focused on inter-species distances and this is why we disabled the automatic adjustment of sample scores in the *Diagram Settings* dialog box.

The separation among the classes is best seen if the points corresponding to different classes are shown with different symbol types and if an envelope encloses all the points belonging to a particular class. In addition, we do not need to have individual observations labelled.

Start with the *Project > Classify > Samples* command. Click the *New from data* button and select all four dummy variables in the *Species* category and check the *Combine dummy variables* option at the dialog bottom. Click the *Create* button and confirm the resulting classification with the *OK* button in the next

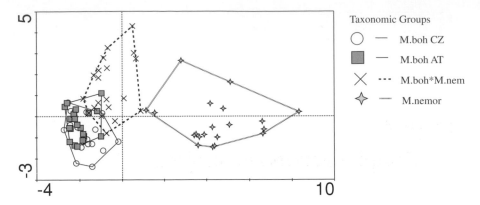

Figure 17-3. Canonical scores of individual plants on first two discriminating axes.

dialog. When you return to the *Available Classifications* dialog, check the *Use this classification ...* option and *Close* the dialog.

Now you should go into the *Project > Settings* dialog and check the *Draw envelopes around Classes of – Samples* option in the *Contents* page. Then move to the *Appearance* page and select the *None* option in the *Labelling of scores > Sample labels*. You can also ask for creation of a diagram legend in this page.

After you close this dialog, execute the *Create > Scatter Plots > Samples* command. The resulting diagram is shown in Figure 17-3.

Note that the two groups representing *M. bohemicum* populations collected in two countries overlap substantially, while the hybrid plants overlap only a little with the *M. bohemicum* parent and there is no overlap of the *M. nemorosum* group with any of the other groups. The diagram is adjusted by differentiating line styles for group envelopes and the filling and symbol type options, to make the graph easier to read from a grey-scale output. Additionally, the group names are slightly edited.

Appendix A:

Sample datasets and projects

The sample datasets, Canoco and CanoDraw project files, and the files with results of the analyses discussed in this book are available from our web site: http://regent.bf.jcu.cz/maed/

You can download the sample files from there and find additional information, including errata and additional notes about the topics discussed in this book.

The sample files are available in compressed ZIP files, either separately for each chapter or in one large ZIP file, containing all the files.

Appendix B:

Vocabulary

Terms in bold text are listed as separate entries in this Appendix.

Attribute plot — Scatter diagram based on sample **scores** and displaying a particular attribute of samples (values of selected **species**, selected **environmental variable**, diversity of species composition etc.)

Biplot — **Ordination diagram** displaying (at least) two different types of objects (e.g. **species** and **samples** or **samples** and **environmental variables**) that can be jointly interpreted using the biplot rule (see Section 10.1)

Canonical axis, Constrained axis — Axis of the ordination space which is constrained to be defined by a linear multiple regression model using the selected **environmental variables** as predictors

Centring — Change of values of a variable, respecting particular context; for example, all the values of a particular variable are changed using a shared transformation 'recipe' to achieve zero average value. Centring is, in contrast to **standardization**, performed by a subtraction (or addition) operation

Covariables — Also named *covariates* in other textbooks or statistical software. These are **explanatory variables** with their effects upon the response variables (**species** data) acknowledged, but not of primary interest. The covariables are used in *partial ordination analyses*, where their effect is partialled out of the ordination method solution

Degrees of freedom — Measure of complexity ('number of parameters') of a statistical model. DFs are also used to express amount of information left in the statistical sample after information has been extracted with a model

Dummy variable	A variable with 0 and 1 values (most often), coding a particular level of a **nominal** (factor) **variable**
Eigenvalue	Measures the importance of the ordination axis, expressed as the amount of variability in the primary data (response variables, **species** data) explained by the corresponding axis. This is a specific interpretation of this term, which has a more general meaning in the matrix algebra
Environmental variables	These are the **explanatory variables** in the ordination methods that are of primary interest for interpretation. The other kind of **explanatory variable** is a **covariable**. A particular variable might act as the *environmental variable* in one analysis and as the *covariable* in another one
Explanatory variables	Variables used to explain directly (in constrained gradient analysis) or indirectly the variability in the primary (**species**) data. Depending on their semantic role in the particular analysis, they are further classified either as **environmental variables** or as **covariables**
Gradient analysis	A (mostly) multivariate statistical method which attempts to explain the abundances of **species** (or, more generally, of any response variables) using continuous (quantitative) **explanatory variables**. In the typical context, these variables are supposed to correspond to a variation of the environment, but they do not need to be directly measured (e.g. in indirect ordination methods). The position of individual **samples** is determined along the calculated gradients. **Ordination** methods are a specific group of gradient analysis methods (see Chapter 3)
Joint plot	**Ordination diagram** displaying (at least) two different types of objects (e.g. **species** and **samples**) that are jointly interpreted by the centroid principle and not by the biplot principle (see Section 10.2)
Linear ordination methods	Ordination methods based on a **species response model**, where the straight lines are supposed to best describe the change of the **species** values along the ordination axes
Nominal variable (=factor)	Variable which has a non-quantitative character and indicates exclusive classes to which individual observations belong

Ordination diagram	Two-dimensional scatter diagram displaying, using symbols and/or arrows, **scores** of objects of one or several types (e.g. **species, samples, environmental variables**), enabling an easier interpretation of the multivariate data sets, summarized by an ordination method. An important property of the ordination diagram is that its two axes cannot be arbitrarily rescaled in respect of one to the other
Permutation test	Test of null hypothesis, where the obtained value of the test statistic is compared with a constructed estimate of its distribution under the validity of a null hypothesis. This distribution is achieved by randomly permuting ('shuffling') the sampling units (observations) in accordance with the particular null hypothesis
Relevé	Traditional term from vegetation science. It refers to a **sample** describing vegetation composition (usually percentage cover of individual plant species or estimate of this cover on a semi-quantitative scale, e.g. the Braun–Blanquet scale)
Sample	In this book, we use the term **sample** differently from its usual statistical meaning (for simplicity and to keep its use in line with the documentation for the CANOCO software). We mean by *sample* one sampling unit (object, site, one row in the data matrix)
Scores	These values represent coordinates of various types of entities (**sample** scores, **species** scores, scores of **environmental variables**) in the ordination space, calculated by the ordination method. They are used to create **ordination diagrams**
SD units	Represent the extent of variation along ordination axes constructed by unimodal ordination methods. Correspond to weighted standard deviation units
Species	We refer by this term to the response (dependent) variables in ordination methods (i.e. the columns in the primary data matrix, the **species** data matrix)
Species response model	Refers to the fitted curve/line describing the assumed shape of the change of values of the response variable (a population size or percentage cover of a biological species, for example) with the values of **explanatory variables** (including here even hypothetical gradients, such as ordination axes from unconstrained **gradient analysis**)

Standardization	Change of values of a variable, respecting a particular context; for example, all the values of a particular variable are changed using a shared transformation 'recipe' to achieve unit variation or unit sum of the values. Standardization is, in contrast to **centring**, performed by a multiplication or by a division operation
Supplementary species, samples, env. variables	The **species, samples,** or **environmental variable** that were not used during the calculations performed by the ordination method. But based on the ordination results, we can project them passively into the ordination space. Therefore, these entities do not contribute to the meaning of the ordination results, but their meaning is interpreted using those results
Transformation	Transformation is, generally, a change of values of a variable, using a parametric (functional) 'recipe'. In this book, we refer to transformation when the individual entries in a data table are transformed with an identical recipe, not varying across samples or variables. Most often, monotonous transformations are used (which do not change ordering of transformed values), such as log or square root.
Triplot	An **ordination diagram** containing three types of objects (**species, samples,** and **environmental variables,** in most cases) where all three possible pairs can be interpreted using the biplot rule (see also the definition of the **biplot** term)
Unimodal methods, Weighted averaging methods	Ordination methods based on a **species response model**, where a symmetrical, unimodal (bell-shaped) curve is assumed to best describe the change of the **species** values along the ordination axes

Appendix C:

Overview of available software

The use of any multivariate statistical method even for small datasets requires a computer program to perform the analysis. Most of the known statistical methods are implemented in several statistical packages. In this book, we demonstrate how to use the ordination methods with possibly the most widely employed package, Canoco for Windows. We also show how to use the methods not available in CANOCO (clustering, NMDS, ANOVA) with the general package Statistica for Windows. In the next paragraph, we provide information on obtaining a trial version of the CANOCO program, which you can use to work through the tutorials provided in this book, using the sample datasets (see Appendix A for information on how to obtain the datasets). We also provide an overview of other available software in tabular form and show both the freely available as well as the commercial software. The attention is focused on the specialized software packages, targeting ecologists (or biologists), so we do not cover general statistical packages such as S-Plus, SAS or GENSTAT.

The Canoco for Windows program is commercial software requiring a valid licence for its use. But we reached agreement with its distributor (Microcomputer Power, Ithaca, NY, USA), who will provide you on request with a trial version of the software, which will be functional for a **minimum** of one month. You can use it to try the sample analyses discussed in this book, using the data and CANOCO projects provided on our web site (see Appendix A). To contact Dr Richard Furnas from Microcomputer Power, write to the following E-mail address: *trial@microcomputerpower.com*.

Table 20-1 lists the main available packages, which can be used for some or many parts of the multivariate analysis of ecological data, and their functionality is compared. We do not own all the listed programs, so the information is often based on the data excerpted from their web pages. For most of them, the authors or distributors reviewed and corrected the provided information.

Table 20-1. Overview of the functionality, relevant to the book contents, of the most often used commercial and freely available software

	ADE4	CAP	Canoco	DECODA	ECOM	MVSP	NTSys-pc	PATN	PCORD	Primer	R	SYN-TAX	Vegan / R	ViSTA
Distribution[1]	F	C	C	C	C	C	C	C	C	C	F	C	F	F
Mac[2]	yes	no	no	no	no	no	no	no	no	no	yes	yes	yes	yes
transf[3]	yes	yes	yes	yes	yes	yes	yes	yes	yes	yes	yes	yes	no[9]	?
stand[3]	yes	yes	yes	yes	yes	yes	yes	yes	yes	yes	yes	yes	yes	yes
PCA	yes	yes	yes	no	no	yes	yes	yes	yes	yes	yes	yes	no[9]	yes
RDA	yes	no	yes	no	yes	no	no	no	no	no	yes	yes	no	yes
CA	yes	yes	yes	yes	no	yes	yes	yes	yes	no	yes	yes	yes	yes
DCA	no	yes	yes	yes	no	yes	no	yes	yes	no	no	no	yes	no
CCA	yes	yes	yes	yes	yes	yes	no	no	yes	no	yes	yes	yes	no
perm. tests[4]	yes	no	yes	no	yes	no	no	yes	yes	no	yes	no	yes	no
partial ord[5]	yes	no	yes	no	no	no	no	no	no	no	yes	no	yes	no
CVA	yes	no	yes	no	no	no	yes	no	no	no	yes	yes	no	no
PCoA	yes	no	yes	?	no	yes	yes	yes	no	no	yes	yes	no[9]	yes
NMDS	no	yes	no	yes	no	no	yes	yes	yes	yes	no	yes	yes	yes
Mantel test	yes	no	no	?	no	no	yes	yes	yes	yes	yes	no	yes	no
clustering[6]	yes	yes	no	several	no	yes	yes	yes	yes	yes	yes	yes	no[9]	yes
coefficients[7]	many	few	several	several	no	many	many	many	many	few	many	many	several	many
TWINSPAN	no	yes	no	yes	no	no	no	yes	yes	no	no	no	no	no
regression[8]	yes	no	yes	no?	yes	no	yes	yes	no	no	yes	no	no[9]	yes

[1] **F** – the software is provided for free, **C** – commercial software;

[2] **yes** means the program runs natively on **Mac**Intosh platforms: where **no** is shown, the program can probably be executed using one of the commercial emulator programs available from third parties; all the programs run on Microsoft Windows platform;

[3] **transf** – basic transformations (such as log or sqrt) are provided, **stand** – standardizations by data columns and rows are available;

[4] permutation tests refer here only to permutation testing of multivariate hypotheses in the framework of RDA/CCA ordination methods, as discussed in this book (but see also the Mantel test further down in the table);

[5] **yes** means that the program uses the concept of covariates (covariables);

[6] here we refer to hierarchical agglomerative **clustering**, many packages also implement the other methods;

[7] does the package provide calculation of distance (dissimilarity) or similarity coefficients in the appropriate context (such as PCoA, NMDS, clustering);

[8] multiple linear regression;

[9] this functionality is provided by the base R system, see Vegan details in the list at the end of this Appendix.

The lists of features are actual as of July 2002 and might not be fully correct. The table is followed by footnotes explaining the meaning of individual rows. Then we provide individual paragraphs for each of the listed programs, with contact information for the distributing companies or individuals and also additional comments, when needed. Five additional programs (CLUSTAN, DistPCoA, NPMANOVA, PolynomialRdaCca, and RdaCca) are listed only there, as their functionality is rather specialized. Question marks in Table 20-1 imply that we do not know about this particular aspect of program functionality.

There now follows the list of contact information for the individual producers of the software listed in Table 20-1. The information provided was accurate at the time of printing.

- **ADE-4**, is available from server of Lyon University <http://pbil.univ-lyon1. fr/ADE-4/>
- **CANOCO for Windows**, version 4.5, is distributed by Microcomputer Power, USA, <http://www.microcomputerpower.com> and by Scientia, Hungary <http://ramet.elte.hu/~scientia/>
- **CAP**, version 2.0, is distributed by Pisces Conservation Ltd., UK <http://www.pisces-conservation.com/>
- **DECODA**, version 2.05, is distributed by Anutech Pty. Ltd., Australia <http://www.anutech.com.au/TD/DECODA_WWW/welcome.html>
- **ECOM**, version 1.33, is distributed by Pisces Conservation Ltd., UK <http://www.pisces-conservation.com/>
- **MVSP**, version 3, is distributed by Kovach Computing Services, UK <http://www.kovcomp.co.uk/mvsp/index.html>
- **NTSYSpc**, version 2.1, is distributed by Exeter Software, USA <http://www.exetersoftware.com/cat/ntsyspc.html>
- **PATN** is distributed by CSIRO, Australia <http://www.cse.csiro.au/CDG/PATN/>
- **PCORD**, version 4, is distributed by MjM Software, USA <http://www.pcord.com>
- **PRIMER**, version 5, is distributed by Primer-E Ltd., UK <http://www.primer-e.com>
- The **R package**, version 4, is available from Philippe Casgrain web site <http://www.fas.umontreal.ca/BIOL/casgrain/en/labo/R/v4/>. Please, note that this software's name can be confused with the more general statistical package R, which is a non-commercial version of the S / S-Plus system (see also the Vegan package below).
- **SYN-TAX**, version 2000, is available from Scientia, Hungary <http://ramet.elte.hu/~scientia/>

- **Vegan**, version 1.5.2, is an add-on package for the R statistical system
  (<http://cran.r-project.org/>).
  Vegan package is available from Jari Oksanen,
  <http://cc.oulu.fi/~jarioksa/softhelp/vegan.html>
- **ViSTA**, version 6.4, is available from Prof. Forrest Young web site
  <http://forrest.psych.unc.edu/research/>
- **CLUSTAN**, version 5.0, is a package specialized in clustering and it
  provides almost every known algorithm of hierarchical agglomerative
  clustering (in addition to other methods such as K-means clustering) and
  calculation of many coefficients of (dis)similarity. It is available from the
  Clustan company, UK <http://www.clustan.com>
- **DistPCoA** is a simple program calculating principal coordinates analysis
  (PCoA) with optional correction for negative eigenvalues; it is also able to
  calculate the starting matrix of distances, using one of several available
  distance measures. It is available from P. Legendre's web site, at
  <http://www.fas.umontreal.ca/biol/casgrain/en/labo/distpcoa.html>
- **NPMANOVA** is a program written by M.J. Anderson, and it calculates
  a non-parametric multivariate ANOVA, based on a selected kind of
  distance (dissimilarity) measure. It supports the two-way balanced
  ANOVA designs only. It is available, together with other programs, at
  <http://www.stat.auckland.ac.nz/~mja/Programs.htm>
- **PolynomialRdaCca** is a simple program calculating both polynomial
  and linear versions of redundancy analysis or canonical correspondence
  analysis. It is available from P. Legendre's web site, at
  <http://www.fas.umontreal.ca/biol/casgrain/en/labo/plrdacca.html>
- **RdaCca** is a simple program calculating redundancy analysis or canonical
  correspondence analysis. It is available from P. Legendre's web site, at
  <http://www.fas.umontreal.ca/biol/casgrain/en/labo/rdacca.html>

References

Anderson, M.J. & Ter Braak, C.J.F. (2002): Permutation tests for multi-factorial analysis of variance. *Journal of Statistical Computation and Simulation* (in press)

Batterbee, R.W. (1984): Diatom analysis and the acidification of lakes. *Philosophical Transactions of the Royal Society of London. Series B: Biological Sciences (London)*, **305**: 451–477

Birks, H.J.B. (1995): Quantitative palaeoenvironmental reconstructions. In: Maddy, D. & Brew, J.S. [eds] *Statistical Modelling of Quaternary Science Data, Technical Guide 5.* Cambridge, Quaternary Research Association: pp. 161–254

Birks, H.J.B., Peglar S.M. & Austin H.A. (1996): An annotated bibliography of canonical correspondence analysis and related constrained ordination methods 1986–1993. *Abstracta Botanica*, **20**: 17–36

Birks, H.J.B., Indrevær, N.E. & Rygh, C. (1998): An annotated bibliography of canonical correspondence analysis and related constrained ordination methods 1994–1996. Bergen, Norway: Botanical Institute, University of Bergen. 61 pp.

Borcard, D., Legendre, P. & Drapeau, P. (1992): Partialling out the spatial component of ecological variation. *Ecology*, **73**: 1045–1055

Bray, R.J. & Curtis, J.T. (1957): An ordination of the upland forest communities of southern Wisconsin. *Ecological Monographs*, **27**: 325–349

Cabin, R.J. & Mitchell, R.J. (2000): To Bonferroni or not to Bonferroni: when and how are the questions. *Bulletin of the Ecological Society of America*, **81**: 246–248

Chambers, J.M. & Hastie, T.J. (1992): *Statistical Models in S.* Pacific Grove: Wadsworth & Brooks.

Cleveland, W.S. & Devlin, S.J. (1988): Locally-weighted regression: an approach to regression analysis by local fitting. *Journal of the American Statistical Association*, **83**: 597–610

Cox, T.F. & Cox, M.A.A. (1994): *Multidimensional Scaling.* London: Chapman & Hall.

Diamond, J. (1986): Overview: laboratory experiments, field experiments, and natural experiments. In: Diamond, J. & Case, T.J. [eds.] *Community Ecology.* New York: Harper & Row, pp. 3–22.

Ellenberg, H. (1991): Zeigerwerte von Pflanzen in Mitteleuropa. *Scripta Geobotanica*, **18**: 1–248

Eubank, R.L. (1988): *Smoothing Splines and Parametric Regression.* New York: Marcel Dekker.

Flinders, C.A. & Magoulick, D. D. (2002): Partitioning variance in lotic crayfish community structure based on a spatial scale hierarchy. (Manuscript submitted)

Gower J.C. (1966): Some distance properties of latent root and vector methods used in multivariate analysis. *Biometrika* **53**: 325–338

Gower, J.C. & Legendre, P. (1986): Metric and Euclidean properties of dissimilarity coefficients. *Journal of Classification*, **3**: 5–48

Grassle, J.F. & Smith, W. (1976): A similarity measure sensitive to the contribution of rare species and its use in investigation of variation in marine benthic communities. *Oecologia*, **25**: 13–22

Green, R.H. (1979): *Sampling Design and Statistical Methods for Environmental Biologists*. New York: J. Wiley.

Grime, J.P., Hodgson, J.G. & Hunt, R. (1988): *Comparative Plant Ecology*. London: Unwin Hyman.

Grubb, P.J. (1977): The maintenance of species-richness in plant communities: the importance of the regeneration niche. *Biological Reviews*, **52**: 107–145

Hájek, M., Hekera, P. & Hájková, P. (2002): Spring fen vegetation and water chemistry in the Western Carpathian flysch zone. *Folia Geobotanica*, **37**: 205–224

Hallgren, E., Palmer, M.W. & Milberg, P. (1999): Data diving with cross-validation: an investigation of broad-scale gradients in Swedish weed communities. *Journal of Ecology*, **87**: 1037–1051

Hastie, T.J. & Tibshirani, R.J. (1990): *Generalized Additive Models*. London: Chapman and Hall, 335 pp.

Hennekens, S.M. & Schaminee, J.H.J. (2001): TURBOVEG, a comprehensive data base management system for vegetation data. *Journal of Vegetation Science*, **12**: 589–591

Hill, M.O. (1979): *TWINSPAN – a FORTRAN Program for Arranging Multivariate Data in an Ordered Two-way Table by Classification of the Individuals and Attributes*. Ithaca: Section of Ecology and Systematics, Cornell University.

Hill, M.O. & Gauch, H.G. (1980): Detrended correspondence analysis, an improved ordination technique. *Vegetatio*, **42**: 47–58

Hill, M.O., Bunce, R.G.H. & Shaw, M.V. (1975): Indicator species analysis, a divisive polythetic method of classification, and its application to survey of native pinewoods in Scotland. *Journal of Ecology*, **63**: 597–613

Hotelling, H. (1933): Analysis of a complex of statistical variables into principal components. *Journal of Educational Psychology*, **24**: 417–441, 498–520

Hurlbert, S.H. (1984): Pseudoreplication and the design of ecological field experiments. *Ecological Monographs*, **54**: 187–211

Hutchinson, G.E. (1957): Concluding remarks. *Cold Spring Harbor Symposia on Quantitative Biology*, **22**: 415–427

Jaccard, P. (1901): Etude comparative de la distribution florale dans une portion des Alpes et du Jura. *Bulletin de la Société Vaudoise des Sciences Naturelles*, **37**: 547–579

Jackson, D.A. (1993): Stopping rules in principal components analysis: a comparison of heuristical and statistical approaches. *Ecology*, **74**: 2204–2214

Jongman, R.H.G., Ter Braak, C.J.F. & Van Tongeren, O.F.R. (1987): *Data Analysis in Community and Landscape Ecology*. Wageningen, The Netherlands: Pudoc. Reissued in 1995 by Cambridge University Press.

Knox, R.G. (1989): Effects of detrending and rescaling on correspondence analysis: solution stability and accuracy. *Vegetatio*, **83**: 129–136

Kovář, P. & Lepš, J. (1986): Ruderal communities of the railway station Ceska Trebova (Eastern Bohemia, Czechoslovakia) – remarks on the application of classical and numerical methods of classification. *Preslia*, **58**: 141–163

Kruskal, J.B. (1964): Nonmetric multidimensional scaling: a numerical method. *Psychometrika*, **29**: 115–129

Legendre, P. & Anderson, M.J. (1999): Distance-based redundancy analysis: testing multi-species responses in multi-factorial ecological experiments. *Ecological Monographs*, **69**: 1–24

Legendre, P. & Gallagher, E.D. (2001): Ecologically meaningful transformations for ordination of species data. *Oecologia*, **129**: 271–280

Legendre, P. & Legendre, L. (1998): *Numerical Ecology*, second English edition. Amsterdam: Elsevier Science B.V.

Lepš, J. (1999): Nutrient status, disturbance and competition: an experimental test of relationship in a wet meadow. *Journal of Vegetation Science*, **10**: 219–230

Lepš, J. & Buriánek, V. (1990): Pattern of interspecific associations in old field succession. In: Krahulec, F., Agnew, A.D.Q., Agnew, S. & Willems, J.H. [eds] *Spatial Processes in Plant Communities*. The Hague: SPB Publishers, pp. 13–22

Lepš, J. & Hadincová, V. (1992): How reliable are our vegetation analyses? *Journal of Vegetation Science*, **3**: 119–124

Lepš, J., Prach, K. & Slavíková, J. (1985): Vegetation analysis along the elevation gradient in the Nízké Tatry Mountains (Central Slovakia). *Preslia*, **57**: 299–312

Lepš, J., Novotný, V. & Basset, Y. (2001): Habitat and successional status of plants in relation to the communities of their leaf-chewing herbivores in Papua New Guinea. *Journal of Ecology*, **89**: 186–199.

Lindsey, J.K. (1993): *Models for Repeated Measurement*. Oxford: Oxford University Press.

Little, R.J.A. & Rubin, D.B. (1987): *Statistical Analysis with Missing Data*. New York: J. Wiley.

Ludwig, J.A. & Reynolds, J.F. (1988): *Statistical Ecology*. New York: J. Wiley.

Mahalanobis, P.C. (1936): On the generalized distance in statistics. *Proceedings of the National Institute of Science, India*, **2**: 49–55

McCullagh, P. & Nelder, J.A. (1989): *Generalized Linear Models*, second edition. London: Chapman and Hall, 511 pp.

McCune, B. & Mefford, M.J. (1999): *PC-ORD. Multivariate Analysis of Ecological Data*, version 4. Gleneden Beach, Oregon, USA: MjM Software Design.

Moravec, J. (1973): The determination of the minimal area of phytocenoses. *Folia Geobotanica & Phytotaxonomica*, **8**: 429–434

Morisita, M. (1959): Measuring of interspecific association and similarity between communities. *Memoirs of the Faculty of Science (Kyushu University), Series E*, **3**: 65–80

Mueller-Dombois, D. & Ellenberg, H. (1974): *Aims and Methods of Vegetation Ecology*. New York: J. Wiley.

Novotný, V. & Basset, Y. (2000): Rare species in communities of tropical insect herbivores: pondering the mystery of singletons. *Oikos*, **89**: 564–572

Okland, R.H. (1999): On the variation explained by ordination and constrained ordination axes. *Journal of Vegetation Science*, **10**: 131–136

Oksanen, J. & Minchin, P. (1997): Instability of ordination results under changes in input data order: explanations and remedies. *Journal of Vegetation Science*, **8**: 447–454

Orloci, L. (1967): An agglomerative method for classification of plant communities. *Journal of Ecology*, **55**: 193–205

Orloci, L. (1978): *Multivariate Analysis in Vegetation Research*, second edition. The Hague: W. Junk B. V.

Pielou, E.C. (1977): *Mathematical Ecology*. New York: J. Wiley.

Pyšek, P. & Lepš, J. (1991): Response of a weed community to nitrogen fertilization: a multivariate analysis. *Journal of Vegetation Science*, **2**: 237–244

Reckhow, K.H. (1990): Bayesian-inference in non-replicated ecological studies. *Ecology*, **71**: 2053–2059

Rice, W.R. (1989): Analyzing tables of statistical tests. *Evolution*, **43**: 223–225

Robinson, P.M. (1973): Generalized canonical analysis for time series. *Journal of Multivariate Analysis*, **3**: 141–160

Shepard, R.N. (1962): The analysis of proximities: multidimensional scaling with an unknown distance function. *Psychometrika*, **27**: 125–139

Silvertown, J. & Dodd, M. (1996): Comparison of plants and connecting traits. *Philosophical Transactions of the Royal Society of London. Series B: Biological Sciences (London)*, **351**: 1233–1239

Sneath, P.H.A. (1966): A comparison of different clustering methods as applied to randomly-spaced points. *Classification Society Bulletin*, **1**: 2–18

Sokal, R.R. & Rohlf, F.J. (1995): *Biometry*, third edition. New York: W.H. Freeman, 887 pp.

Sörensen, T. (1948): A method of establishing groups of equal amplitude in plant sociology based on similarity of species contents and its application to analysis of the vegetation on Danish commons. *Biologiska Skrifter (Copenhagen)*, **5**: 1–34

Špačková, I., Kotorová, I. & Lepš, J. (1998): Sensitivity of seedling recruitment to moss, litter and dominant removal in an oligotrophic wet meadow. *Folia Geobotanica*, **33**: 17–30

Štech, M. (1998): Variability of selected characteristics of the species from section *Laxiflora (WETTSTEIN) SOÓ 1927* and revision of the genus *Melampyrum L.* in the Czech Republic. PhD thesis [in Czech], University of South Bohemia, C. Budejovice, Czech Republic

Stewart-Oaten, A., Murdoch, W.W. & Parker, K.P. (1986) Environmental impact assessment: 'pseudoreplication' in time? *Ecology*, **67**: 929–940

Ter Braak, C.J.F. (1994): Canonical community ordination. Part I: basic theory and linear methods. *Ecoscience*, **1**: 127–140

Ter Braak, C.J.F. & Prentice, I.C. (1988): A theory of gradient analysis. *Advances in Ecological Research*, **18**: 93–138

Ter Braak, C.J.F. & Šmilauer, P. (2002): *CANOCO Reference Manual and CanoDraw for Windows User's Guide: Software for Canonical Community Ordination (version 4.5)*. Ithaca, NY: Microcomputer Power, 500 pp.

Ter Braak, C.J.F. & Verdonschot, P.F.M. (1995): Canonical correspondence analysis and related multivariate methods in aquatic ecology. *Aquatic Sciences*, **57**: 255–289

Underwood, A.J. (1997): *Experiments in Ecology*. Cambridge: Cambridge University Press, 504 pp.

Van den Brink, P.J. & Ter Braak, C.J.F. (1998): Multivariate analysis of stress in experimental ecosystems by Principal Response Curves and similarity analysis. *Aquatic Ecology*, **32**: 163–178

Van den Brink, P.J. & Ter Braak, C.J.F. (1999): Principal Response Curves: Analysis of time-dependent multivariate responses of a biological community to stress. *Environmental Toxicology and Chemistry*, **18**: 138–148

Van der Maarel, E. (1979): Transformation of cover-abundance values in phytosociology and its effect on community similarity. *Vegetatio*, **38**: 97–114

Von Ende, C.N. (1993): Repeated measures analysis: growth and other time-dependent measures. In: Scheiner, S.M. & Gurevitch, J. [eds] *Design and Analysis of Ecological Experiments*, New York: Chapman and Hall, pp. 113–117

Wartenberg, D., Ferson, S. & Rohlf, F.J. (1987): Putting things in order: a critique of detrended correspondence analysis. *American Naturalist*, **129**: 434–448

Whittaker, R.H. (1967): Gradient analysis of vegetation. *Biological Review (Camb.)*, **42**: 207–264

Whittaker, R.H. (1975): *Communities and Ecosystems*. New York: MacMillan.

Williams, W.T. & Lambert, J.M. (1959): Multivariate methods in plant ecology. I. Association analysis in plant communities. *Journal of Ecology*, **49**: 717–729

Wollenberg, A.L. van den (1977): Redundancy analysis. An alternative for canonical correlation analysis. *Psychometrika*, **42**: 207–219

Wright, S.P. (1992): Adjusted P-values for simultaneous inference. *Biometrics*, **48**: 1005–1013

Index